Human Microbes - The Power Within

Vasu D. Appanna

Human Microbes - The Power Within

Health, Healing and Beyond

Springer

Vasu D. Appanna
Department of Chemistry and Biochemistry, Faculty of Science and Engineering
Laurentian University
Sudbury, Ontario
Canada

ISBN 978-981-10-7683-1 ISBN 978-981-10-7684-8 (eBook)
https://doi.org/10.1007/978-981-10-7684-8

Library of Congress Control Number: 2018932163

Printed on acid-free paper

This Springer imprint is published by Springer Nature
The registered company is Springer Nature Singapore Pte Ltd.
The registered company address is: 152 Beach Road, #21-01/04 Gateway East, Singapore 189721, Singapore

Preface

This book is a journey into one of the human organ that has long been ignored due to its invisibility and dispersed nature. However, technological advances are now enabling the visualization of the microbiome in its undisturbed environment where its various constituents can be seen diligently shepherding the human body to function effectively. All the microbes including the viruses, bacteria and fungi that live on and within the body are referred to as the microbiome. For the purpose of this book that comprises four chapters bacteria are considered as the main component of this body part invisible to the naked eyes.

Chapter 1 describes the emergence of life on planet earth and posits that bacteria were central in the development of virtually all living organisms. The ability of these minute organisms to live independently and to execute a variety of tasks is critical to the survival of other organisms. Further, the contribution of bacteria to the genesis of the chloroplast and the mitochondrion provided the energy boost needed to sustain life in partnership with the sun. While chloroplast is an integral component of organisms involved in capturing solar energy, mitochondria is part of all multicellular life where they are the main energy generator. These intimate cellular interactions between these organelles which originated from bacteria and living systems provided the fuel life needed to evolve. These transformed bacteria are present in virtually all human cells. Indeed, the trillions of microbes that live on and within the human body make humans the way they are. Only with the recent advent of molecular visualization technologies that this microbe-human link is being fully appreciated. The post-2007 human is someone with a microbial body part weighing 2–3 kg in an average adult that fulfills a variety of essential functions.

Chapter 2 elaborates on how this organ is formed and traces its development from the womb till the old age. The various factors like diet, genetic make-up, geography and life-style that are known to influence the maturation of the visible organs including the lungs, the brain and the heart also dictate the nature of this invisible organ. The microbiome is an expansive organ akin to the blood system. Just as the latter that is constituted by the white blood cells, red blood cells and numerous other components, the microbiome is also composed of disparate microbial cells. These microbes are specific to where they are located in the body and participate in an array of physiological activities that cannot be accomplished in their absence. We will not able to digest numerous food products properly nor produce vital ingredients like vitamins K and amino-acids such as tryptophan without these microbial partners.

Chapter 3 evaluates the abnormal situations that ensue when the microbiome does not function properly. The diseases that manifest following the imbalance amongst the microbial communities are explained. A range of effectors is known to disrupt the fine-balance existing between the visible organs and these diverse microbiota. Unlike the visible organs that are made-up of relatively few specific cells, the functional microbiome is constituted by a large number of disparate microbes. Pollution, diet, exercise, pets, occupation, and hormonal fluctuations are some of the factors known to distress the microbiome. These perturbations can trigger illnesses like gastric ulcer, colorectal cancers, fatty liver disease, hypertension and obesity. The ability of probiotics and prebiotics to re-establish a dysfunctional microbiome and cure diseases is described. Probiotics involve the intake of live-bacteria while prebiotics are more or less like fertilizers aimed at favouring the proliferation of a select group of microbes. These treatments are proving very effective in treating diseases promoted by dysbiosis. These procedures are becoming a common health practice aimed at fostering general wellness.

Chapter 4 provides a glimpse into the future when the molecular functioning of the microbiome will be fully elucidated. If in 10–30 years from now the knowledge regarding this invisible organ has attained the same degree of understanding as the visible organs like the heart, there will be a very momentous societal change. The impact of this information on health education, health delivery systems, modifying human traits, global hunger, and tracking individuals is discussed. For instance, just imagine the presence of a microbiome bank in a hospital and how this will help in patient care; the precise profiling of each individual with the microbes living on and within the body will be a game-changer in medical diagnostics. This will also completely revolutionize molecular forensic identification. The discoveries of unique microbes with distinct characteristics that can ease space travel and help in the production of goods will be of immense economic value. These are some of the fascinating, provocative, and forward-looking ideas discussed in this chapter.

The information synthesized in this book emanates from recent findings on the human microbiome and the possibilities they hold for the future as our understanding on these microbial communities expand exponentially. This discipline is just beginning to emerge as a potent societal game-changer that will affect us all. I am indebted to my wife Dr. Sharina Appanna for assiduously proofreading the original text. I will like to thank all my students who have taught me how to teach and bring complex information to life. I am confident that this attribute is reflected in this book that can be enjoyed by a wide readership including health science students and practitioners.

October 27, 2017

Vasu D. Appanna
Department of Chemistry and Biochemistry
Faculty of Science and Engineering, Laurentian University
Sudbury, Ontario, Canada

Contents

About the Author

Dr. Vasu D. Appanna has been teaching and researching for over 30 years on various aspects of how microbes go about surviving in extreme environments. He is in fact 'a bacterial whisperer' and has been able to 'talk' microbes into executing a variety of challenging tasks. For instance, he has tailored microbes to grow in metal-polluted areas, live on glycerol and adopt a lifestyle on limited oxygen. Dr. Appanna has published in excess of 100 articles in scientific journals and lectured extensively around the globe including France, India, Russia, Chile and China. He is the academic editor of numerous journal publications. He has served as Department Chair and Dean of the Faculty of Science & Engineering at Laurentian University. As a prolific researcher he has trained and mentored over 150 highly qualified research personnel. He has also helped identify some of the mechanisms metal pollutants invoke to trigger obesity and neurological diseases. His passion and knowledge of microbes have taken him around the globe in pursuit of biotechnological solutions. Whether it is the search for microbes to mitigate pollution in the Black Sea or to extend the life of oil reserve in Sarov, Russia, or to improve water quality in the aquaculture industry in Vietnam, or to mine valuable metals in the Atacama desert, Chile, Dr. Appanna has been a pioneer in cajoling microbes in performing a desired chore. In this book he has summoned his extensive teaching skills, microbial expertise and background in human health to weave the fascinating facets of our invisible organ.

Chapter 1
The Human Microbiome: The Origin

Contents

Abstract Microbes are at the origin of all life forms on this planet. They are the first inhabitants, have played and are playing a pivotal role in shaping most living organisms that followed thereafter. The major driving force dictating life is energy. The capture of solar power by the chloroplast, a specialized microbe and the release of energy in the cells by the mitochondrion, another specialized microbe, were central to the emergence of higher organisms. As nature tends to utilize ingredients available in its midst to build, mould and customize its occupants, microbes were a natural choice for this bioengineering task. Hence, microbes became part of most if not all living organisms. They live as symbiotic partners, supply essential nutrients, act as guards and shape morphological and physiological attributes. Although the presence of microbes housed in specific locations in numerous organisms has been detected before, their beneficial interaction with humans has only recently begun to be appreciated. The discovery of new molecular visualization technologies is providing an unprecedented view of the

V.D. Appanna, *Human Microbes - The Power Within*,
https://doi.org/10.1007/978-981-10-7684-8_1

intimate relationship humans have forged with microbial partners. In fact, the vast majority of cells we harbor are microbial in origin. The entirety of microbes living within and on us, referred to as the microbiome, is an integral part of the human body, just like the visible organs such as the heart and the eye. Our understanding of the development and functions of this invisible organ is in its infancy. As more information on this body part is elucidated, its involvement in our well-being and on who we are will become indisputable. Once the molecular details of the human microbiome are unveiled, our lives and our society will change forever.

Keywords Human microbiome · origin · microbe · mitochondrion · chloroplast · energy · symbiosis · fermentation · molecular visualization · invisible organ · indispensable · omnipresent

1.1 The First Colonizers of Planet Earth

Imagine if someone informs you that you are not alone and that your body does not entirely belong to you. That you have inherited only a fraction of your genetic blue-print from your parents and that the overwhelming amount of DNA (the genetic code that goads you along your life adventure) you possess at a given time is not controlled by you. The vast majority of DNA that is located within and on you is found in the microbial communities which are an integral part of you. Scary thought! Well, the majority of the cellular components that you are made up of, i.e. your flesh and blood, is of microbial origin with its own genetic make-up. These microbes constitute an invisible organ that weights nearly two to three kilograms in an adult. They have a lot to say about who you are and how your body ticks (Box 1.1). Only very recently has this invisible part of our body become visible (Fig. 1.1). As we start comprehending this hidden body component, the impact on our health and well-being will be far-reaching. To appreciate this part of ourselves, an invisible organ we have since birth, let us start from the beginning and try to explore how these invisible partners became an inherent component of our existence and how they have shaped who we really are!

Box 1.1 What Makes the Human Tick: The Molecular Tale

Any physiological response and biological activity that the body is engaged in depend both on its visible and invisible components. The information that dictates this reaction is safely tucked in the nucleus of these cells where the chromosomes store the biological commands in the genes. These genes give us our traits (sex, size, colour, etc.) and numerous attributes guiding our day to day living. All the genes that the body possesses are referred to as the genome. A gene is responsible for the generation of the participants that help carry out a specific instruction at a given moment. It is made up of

(continued)

Box 1.1 What Makes the Human Tick: The Molecular Tale (continued)
DNA (deoxyribonucleic acid) where the information is kept in the form of codes, which is essentially locked up in the four bases, adenine, guanine, cytosine and thymine. The commands in the DNA are copied as RNA (ribonucleic acid), an entity also involved in decoding the message in the gene. This results in the production of targeted proteins and metabolites, which are the molecular workers that execute the task demanded by the body. For instance, let us journey through the biological events leading our mouth to water when we smell or see an appetizing food. These sensory signals are perceived by the brain that directs the stomach to secrete gastrin (a hormone, protein). This chemical event activates the gene (DNA) responsible for production of an enzyme (carbonic anhydrase, protein) creating an acidic condition (hydrochloric acid, metabolite) readying the stomach for the arrival of the food, just in case we decide to eat it. However, since the food is still on the dish or on display, the mouth salivates to neutralize the empty acidic stomach (signal-gene-protein-metabolite-biological response). Hence our mouth waters and it takes all these molecular soldiers to trigger this mouth-watering activity (Fig. 1.2).

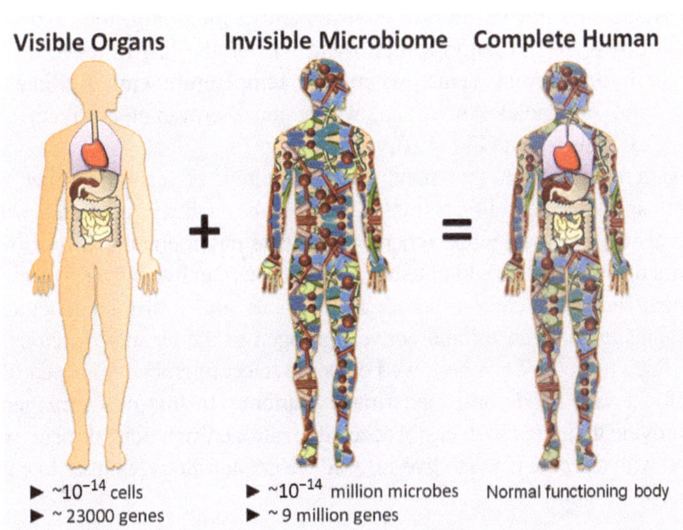

Fig. 1.1 The real you with the visible and invisible constituents. A complete human is made up of all visible organs and all the invisible microbes (bacteria, viruses and fungi) that reside within and in the body. In fact the invisible cells outnumber the visible ones by numerous folds. The proper functioning of the body is dependent on the genetic make-up of all the cellular components, those we can see and cannot see

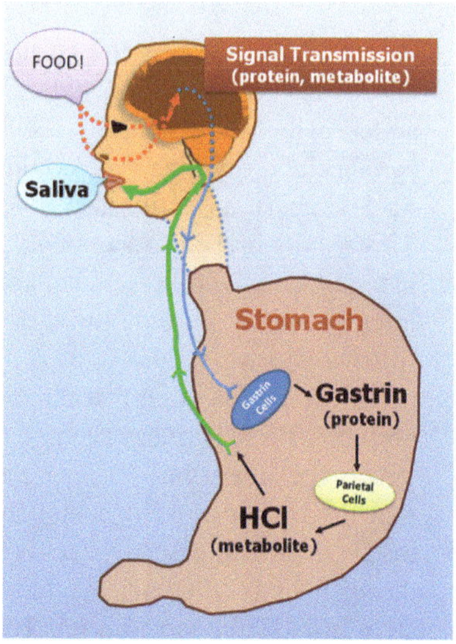

Fig. 1.2 The molecular machines at work. The physiological response the body elicits in the presence of the sight or smell of a delicious food involves the DNA, proteins and metabolites. (signal perceived (DNA, RNA, proteins)-gastrin (protein)-carbonic anhydrase (protein)-hydrochloric acid (metabolic) – salivation (watering-mouth)

1.2 The Beginning: The Diversity of Microbial Life

Microbes were the first inhabitants of this planet and they have learnt to live with all the other residents that came after them, a wise survival trait indeed. They were quick to adapt and thrive in all environments, including the extreme ones. Microbial life was first to colonize earth and will be the last to leave it. They can be found in hydrothermal vents, where the temperature can oscillate between 300–500°C and in glacial Antarctica, where the thermometer hovers around at least −40°C all year round (Fig. 1.3).

They can remain dormant and inanimate for hundreds of years. However, they have the ability to spring back to life. Some of the microbes call areas polluted with metals and chemicals their homes, while other only tolerate environments devoid of oxygen, places where most animals could not survive. Microbes can live on simple nutrients and some of them can utilize carbon dioxide as their sole source of food. They are one of the few organisms that can fix and convert nitrogen in the air into delicious proteins. This feat, which has only been bestowed on some select microbes, has been bartered in exchange for a safe abode and ready-made nutrients. In this neat arrangement, the microbes provide their host with easily accessible nitrogen-rich nourishment, while they are supplied with a secure place to live-in. And we get nutritious legumes to consume.

1.3 The Ubiquitous Microbes: They Are Everywhere

The diversity of habitats tolerated by microbes allows these invisible creatures to affect the lives of all organisms sharing this planet. Irrespective of the environment, one is certain to encounter microbes in and around all organisms, even

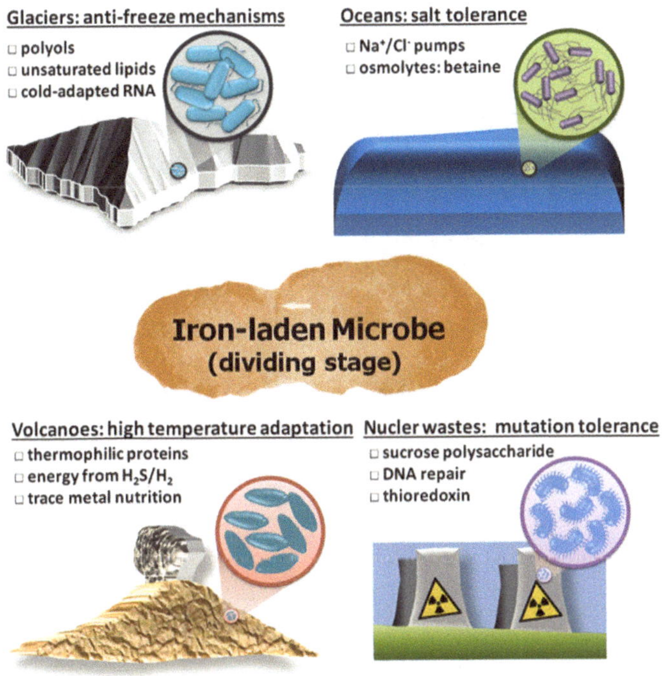

Fig. 1.3 Microbes living in diverse habitats. Microbes are the origin of all living entities on the planet earth, they are present virtually in all ecological niches. A bacterium (*Pseudomonas fluorescens*) in a dividing state laden with Fe (center). They are found in glaciers, in oceans; in volcanoes and in nuclear wastes

though the hosts living in an ecological niche may be very specific to that milieu. For instance, marine organisms are highly adapted to thrive in an aqueous environment, with their gills and other molecular gears to satisfy their demand for oxygen, however, they do have microbes both within their surroundings and on them. The same observation can be made in dry and arid places. In deserts, camels have evolved to confront the lack of water and grazing space available in their landscape. But it goes without saying they are also surrounded by microbes and possess a huge amount these stealthy operators on their skin and within them. A similar situation is evident with plants belonging to the cactus family; these desert-dwellers have perfected their ability to grow in water-deficient environments. They possess a rich collection of microbes that promotes growth and enables them to tolerate drought.

These few illustrations attest to the fact that microbes truly are everywhere and that all organisms have evolved to live with them. It is quite likely that many organisms are morphologically and physiologically the way they are because of the microbes which have accompanied them from the beginning. Hence, it is not surprising that most, if not all occupants of this planet have evolved to flourish in association with these invisible partners (Fig. 1.4).

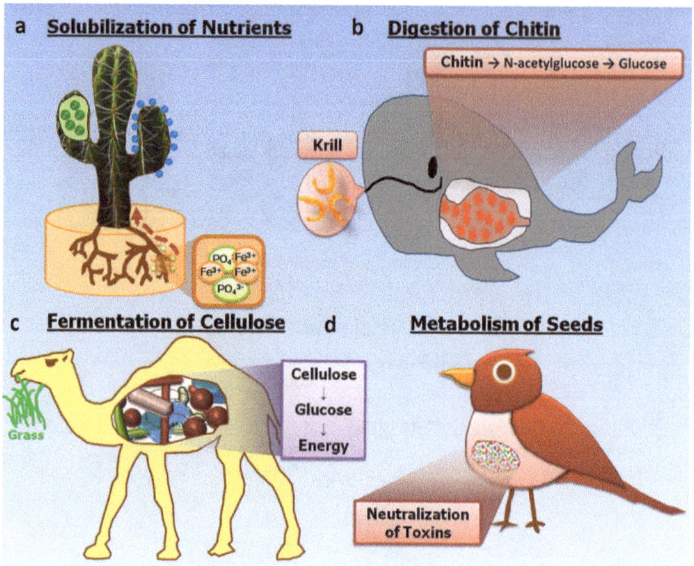

Fig. 1.4 The omnipresent Microbiome. Microbes are an integral component of all multi-cellular organisms and assist them in thriving in a given ecological niche. a Cactus. b Whale. c Camel. d Bird

1.4 Microbes: The Origin of Most Organisms

The first microbes appeared on this planet around 3.6 billion years ago while the first modern human began roaming the earth a mere 200,000 years ago. This huge time gap reflects the meticulous details human evolution has undergone and it is within the realm of possibilities that all these stages may have been steered by the living invisible environment (Fig. 1.5). Then, it stands to reason that as the first occupants of the planet, microbes have contributed to the existence, proliferation, physiological organization and the general well-being of all the organisms which have subsequently followed. Nowhere is this more evident than in the birth of multicellular organisms. The interaction and trapping of one unicellular microbe by another may have triggered this genesis and may have led to the beginning of complex life systems the way we know them today.

The entrapped microbe became the specialized cellular entity referred to as the mitochondrion. The mitochondria have their own genetic information, just like the other confined microbe in plants that is referred to as the chloroplast. Chloroplasts have dedicated their existence to sequestering energy from the sun and transforming it into chemical energy. This is then stored in the form of starch, proteins and lipids found in fruits, vegetables, legumes, leafy greens and other plant-derived treats. Mitochondria on the other hand have become the energy machine par excellence in all aerobic organisms. Any moving or fixed living entity capable of breathing has to have mitochondria. Without the mitochondria, they are doomed to a very lethargic

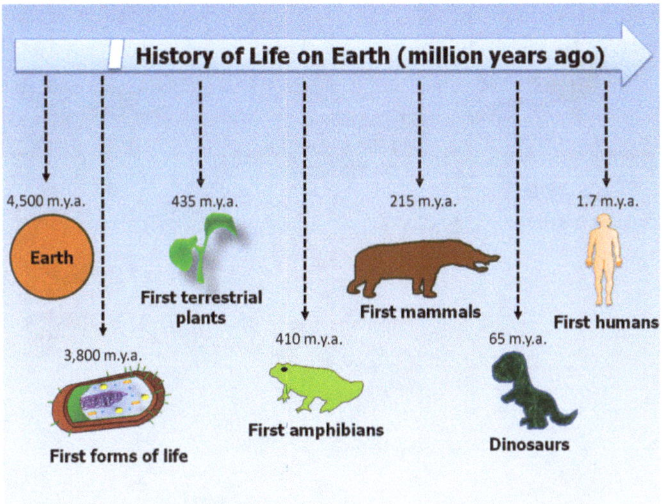

Fig. 1.5 Life on Earth – A chronological history. The emergence of different organisms at various times in the history of planet earth

existence without the energy to maneuver. Hence, it is quite evident that microbes have been at the forefront and most probably acted as the instigator of all life forms we are surrounded by today. The genius of microbes in trapping the near limitless energy generated by sun in the form of chemical power was crucial to propel the evolution of life that we see in our environment. The ability of the other microbial tour de force, aimed at converting this power into a common energy currency, adenosine triphosphate (ATP), was the masterpiece event for the beautiful and radiant life to take its course and blend in the natural landscape at its disposal.

ATP is the universal energy all organisms rely on to perform any cellular task and this practice transcends all known kingdoms of life. Microbes are at the center of it all; sequestering the limitless solar energy and transforming it into a commodity that fuels all biological activities. These events were critical in setting the stage for multicellular life to emerge. The unifying force in a human engaging in any activity, or a bird flying, or a centipede ceremoniously charting its way around the garden or a squirrel scurrying with some dry cones is the omnipresent cellular commodity, ATP. With these specialized microbial siblings, the chloroplast and the mitochondrion fully integrated in different domains of life, microbes were ready to orchestrate and participate in the well-being of most, if not all living organisms. Hence, the evolution of single cell microbial life was indeed the precursor to everything else that came after. The access to abundant energy, which was generated by the chloroplast- mitochondrion duo, propelled the further development of life and led to more cellular sophistication and refinement, which is exhibited in the different life forms we see lurking in our scenic vistas, be it on land, in the air or in the ocean (Fig. 1.6a, b).

The initial magical entrapment of a microbe by another microbe was the beginning of life on this planet the way we know it. This allowed for the efficient

a

b

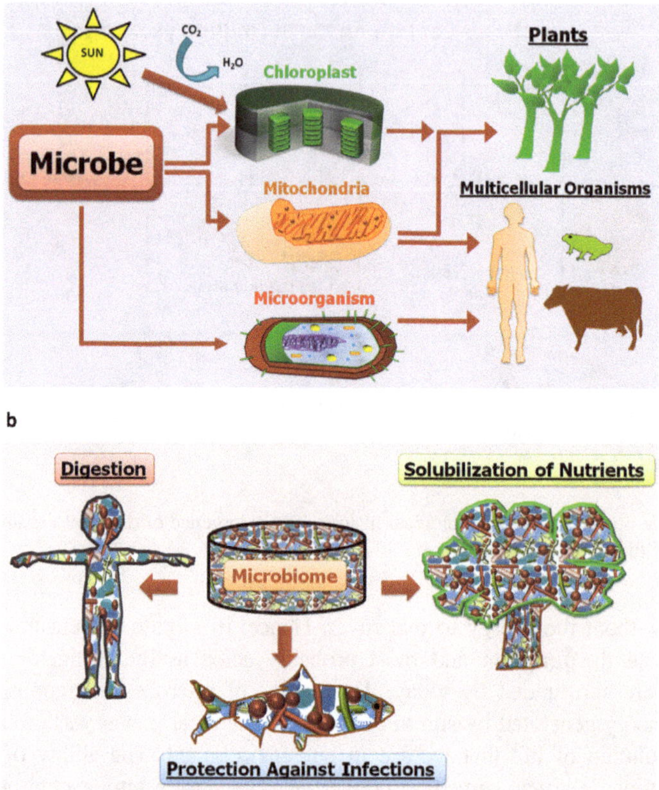

Fig. 1.6 (a) Microbes at the center of life: The energy guru. The ability of microbes to handle the energy from the sun was central to the establishment of life on earth. The chloroplast (a transformed microbe) captures solar energy in photosynthetic systems while the mitochondrion (the other transformed microbe) liberates energy in aerobic organisms. This crucial ingredient fuels all life forms. (b) Microbes set the stage for the emergence of other life forms. With the incorporation of the mitochondria in all O_2 loving organisms, life began its march with the input of the natural elements in environment including microbes. Eukaryotic cell (center); tree, human, fish (various microbes within and on these organisms)

conversion of the inexhaustible solar energy into chemical energy, which triggered the proliferation of different living organisms from the abundance of energy released by the sun. This microbial – solar partnership aimed at the generation of chemical energy was at the center of it all and without these specialized microbial components, life on earth would have taken an entirely different trajectory. Perhaps the exploration of the outer galaxies will reveal some other molecular living secrets, but on earth, the microbe is king and is part and parcel of virtually all living organisms (Box 1.2).

Box 1.2 Microbe: The Harvester and Provider of Energy

The sun is the basis of all the energy on this planet and ATP is the chemical energy that all living organisms utilize in their day to day activity. The ability of the chloroplast (modified microbe adapted to trap the solar energy) to convert sunlight into ATP, which is then utilized to transform carbon dioxide and water into food (starch), was pivotal to the emergence of higher organisms. The energy trapped in the food during photosynthesis needed to be released effectively if other organisms were to emerge. Here, again the microbe, in the form of the mitochondria (microbes dedicated to generating energy in the presence of oxygen residing in all multicellular organisms including humans), provided the proper engine to liberate ATP from the nutrients generated by solar power.

$$\text{Sun} \rightarrow \text{chloroplast} \rightarrow \text{ATP} = \text{food} - \text{mitochondria}$$
$$- \text{ATP higher organisms} \rightarrow \text{humans}$$

The mitochondria supply us with the fuel we need to do our day to day activity and a bit more. In fact, they are the carburetors that are always on the go and literally fired up at all times of the day or night to support us in all our daily activities and shenanigans. In fact, they are on almost from the day we are conceived and will only stop generating ATP on the day we die. During medical conditions, like stroke and cardiac abnormalities, when they cannot function effectively due to the lack of oxygen, we run the risk of permanently damaging our vital organs, including the brain. This critical event, when some microbes lost their independence to become the energy machine, literally provided the necessary fuel boost that gave rise to most life forms As these multicellular living entities begun to emerge, most of them acquired microbial partners to fulfill a variety of daily functions – just try to visualize a cow without its bacteria helping this mammal digest grass. It will be impossible for a cow to have the anatomical structures, like its stomach and the corresponding physiological processes, without input from resident microbial partners. Thus, it stands to reason that we utilize what is available in our ecosystem to not only live but to also to be who we are. As microbes populate our landscape, they have naturally become part and parcel of the development of multicellular life forms, including humans. This microbial relationship with other organisms has continued on unabated, is very dynamic and undergoes continual adjustment. Recent technologies have enabled the mapping of some of the crucial roles our microbial partners play. In fact, they constitute an important body part in the same manner as the heart, brain or liver.

1.5 Microbes in Our Midst: Tale of Our Microbial Friendship

Microbes have been part of our natural landscape, both within us and in our external environment, since the beginning of human life. Although humans were not able to visualize this invisible world until the seventeen century, following the invention of the microscope by the Dutch naturalist, Antonie Van Leeuwenhoek, our ancestors were aware of the beneficial aspects of a variety of the non-discernible elements around them. In fact, the occurrence of this obscure invisible world was first mentioned in the ancient Sanskrit scripture around 1500–1200 BC. The term 'Adrista' was coined to describe the invisible life around us. In all ancient societies the components of this invisible world were often put to good use without the knowledge of what they were. The elements of this unseen world were incorporated in agricultural practices, to treat ailments and to render food more palatable. Rubbing and pressing of soils on wounds were executed in the belief that some magical components within them provided healing power. Studies of fossilized humans have revealed traces of dirt and sediments in teeth and wounds. In ancient China and Greece, moldy bread was utilized to prevent infections. Often, water from a specific lake that was thought to possess divine virtue was used to cure ailments, a ritual enjoyed by many, even today. Microbes and their secretions may have been the architect of these remedies. The earliest known permanent human settlement was in Jericho around 8000–7000 BC. Here agricultural practice was executed in such a way as to naturally enhance the quality of the soil. In this instance, the re-invigoration of fields by rotating cereal farming with leguminous plants was promoted. This custom allowed the soil to be replenished with nutrients and provided a better yield. Indeed, we know now that the microbial life inhabiting these plants was the invisible agents doing their magic in generating a bountiful harvest (Fig. 1.7).

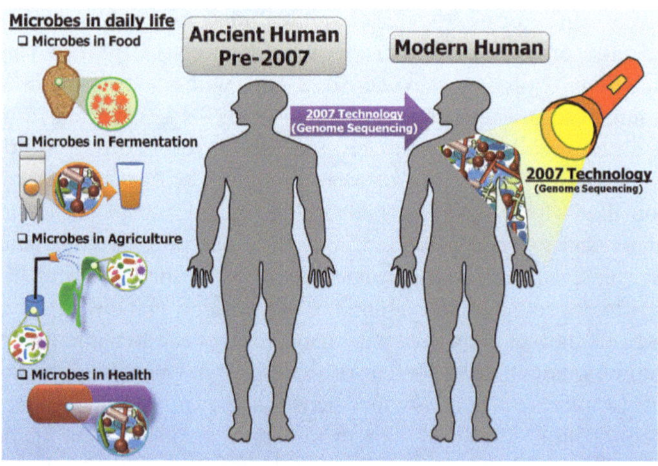

Fig. 1.7 Use of microbes in ancient civilizations. Microbes were an integral part of wine-making; agriculture and curing disease. However their involvement was unknown until the advent of technologies rendering the invisible world visible

1.6 Microbes: Culinary Experts Shaping Our Body

Nowhere have these invisible partners been more at work than in the culinary arena, where they have been busy preserving and transforming foods for our ancestors and for us. Whether we eat microbially-transformed foods or resort to a diet that is rich in starch, proteins or fats, we possess tell-tale molecular signs that reveal our nutritional traditions. In fact, our chewing and tasting machinery has evolved with these microbially processed foods. For instance, one can look no further than the salivary enzyme amylase, as it can reveal our dietary habits. This enzyme, which chops the complex carbohydrate starch into simple glucose, is more prominent in cultures where starch-rich foods are part of the staple diet. Hence, it is more prevalent in the spit of Africans than that of Eskimos, who partake in more protein-rich nutriments. The infusion of microbes in our foods has helped shape our digestive attributes. Even today, the prowess of these invisible cooks is a much sought after expertise in an effort to placate our ever-demanding taste buds. Unbeknownst to them, our forebears were exploiting microbial transformation of raw plant, animal and fish components to produce tongue-twisting flavours that have become a part of our culinary culture.

These nutritionally transformative events were catalyzed spontaneously by the invisible world that was dominant in the diverse geographical regions where humans had settled. Indeed, fermentation became the method of choice to preserve the bounty of nature. Whether it was cabbage in Korea, fish heads in the Arctic or beans in Indonesia the exploit of this invisible world was on display. The omnipresence of microbes in the environment and the nutrient–rich secretions from plant and animal products provided a natural pot for an intimate brewing undertaking to occur. As these invisible food-processors begin to multiply and thrive, they released tiny chemical weapons that enabled them to fend off any competition and suppress other microbes responsible for food spoilage. This is how these invisible cooks toil assiduously to turn cabbage into kimchi, rice-lentil into dosa, grapes into wine and milk into cheese each with a unique flavour and smell that connoisseurs and food buffs revel in. Microbial fermentation was a purely fortuitous occurrence and is presumed to have happened around 7,000–8,000 years ago. Today, it is still used as a powerful preservation technique and food processing tool in all cultures. This welcomed transformation of raw products of nature was a mystery and divine intervention was invoked in the performance of such a feat. Egyptians praised Osiris for brewing beer, while the Greeks were thankful to Bacchus, the god of wine for the magic he performed with grapes. In Japan, the early miso and shoyu breweries had a small shrine that was thought to be the force behind the tasty brew. Workers bowed daily to the shrine and occasional rituals were performed in order to ensure a good brewing outcome (Box 1.3).

Box 1.3 History of Fermentation: Cooking with Microbes
Fermentation may be considered as the oldest food technology that humans have utilized unknowingly. Its origins go back as far as 7000 BC. Cheese and beer were fermented in Assyria. While soya bean was brewed in China, dahi, a fermented milk, was a common drink in ancient India. It was only in 1854 that the microorganisms involved in this process were identified by the French scientist Louis Pasteur. Later, the enzymes involved in the production of lactic acid, acetic acid, ethanol and carbon dioxide, which are key in microbial food processing, were elucidated. The numerous attractive features, like longer shelf life, unique aromas, specific taste, attractive color, characteristic shape and size (remember the blue color cheese with gaping holes) attributed to fermented foods, are due to the microbial production of organic acids, various metabolites, gases and antibiotics. Indeed, these natural products that are attributed to the resident microbes are also touted for their nutritional and medicinal potency.

In Asia the fermentation of vegetables, soya beans and rice was common. Kimchi, a fermented mixture of cabbage, radish and other vegetables permeates all aspects of Korean life even today. The interaction of this invisible world with rice-lentil batter has provided India with dosa and idli, delicious dishes revered by many in South Asia and beyond. Sake, a rice-concocted wine brewed by the microbial brewers is an important part of Japanese tradition, as are dairy products, a signature identity of numerous European countries. For instance, France prides herself on producing unique brands of cheese and yogurt, attributed to specific microbial communities that exist within her geographical boundaries and which are non-existent anywhere else. In fact, the financial stake of these undertakings is so huge that the microbial procedures are protected and offenders are prosecuted if they try to imitate their production. In Africa, the blend of local ingredients and the invisible chefs have given birth to numerous distinct delicious dishes that have become part of the normal diet of the population and that are enjoyed by all. Maize, millet, sorghum and cassava are the most commonly fermented foods throughout Africa. In Kenya uji, a maize product is a regular staple, while in Benin, an uncooked fermented maize dough known as mawe is a national dish. These microbial cooks have become extremely prized items that are tightly protected in order to thwart any attempt by non-native countries to commercialize them (Fig. 1.8).

Hence, it is quite logical to opine that this intimate relationship between the invisible microbial world and the raw materials prevalent in various regions provided a major impetus to the development of human culture, cuisine and physiology. Again, microbes are at the center of these happenings. The raw products that were diligently concocted by these invisible chefs culminated into mouth-watering dishes that our bodies have become accustomed to. This primordial fermentation that our ancestors were also exposed to has contributed to the development of their biological systems. The physiological processes of our ancestors evolved and are evolving with the microbe-laced foods they were and we are consuming. It is well-documented that

Fig. 1.8 Microbes in cooking and preservation in ancient and modern world. Microbes were part of the intimate human landscape in the form of food we consume. a Kimchi (fermented cabbage). b Dosa (fermented lentils). c Mawe (fermented maize). d Cheese (fermented milk). e Idli (fermented rice and lentils)

Koreans who travel without kimchi suffer from poor digestion and experience stomach symptoms, while many Indians are nutritionally–challenged if they are deprived of their dosas. It is fair to conclude that whenever one ventures in a new geographical arena, one's system has to adjust to the invisible microbial world before one can enjoy the local delicacies. The digestive discomfort we experience when we wander into foreign lands is a result of the microbes unknown to our body. Thus, it is not surprising that it was and it is still customary for some travellers to carry their own foods when they depart for a distant land. Again, unknowingly, our gut microbes are dictating how we behave. However, globalization of the foods we consumed may be slowly changing this. The proliferation of fast food chains in remote countries far away from the places where these restaurants originated are to some extent eliminating this traveling ritual, a trend which is transforming our microbiota, and decreasing variety. This disruption of microbial diversity may have an inherent risk and may have a homogenizing influence on our taste-buds, to the chagrin of niche restaurateurs and on our body. The same can be said of our modern lifestyle that is also contributing to this diminishing microbial diversity. In fact, people living in the tribal regions of Tanzania display more varied microbial communities than their urban cousins. The change in dietary habit is indeed a major modulating factor of our microbial partners. This intricate link between the foods that we eat and microbes was all a mystery to our ancestors.

1.7 Unmasking the Invisible World

Despite being out of sight and imperceptible, the microbial world is and has always been shaping the human body. Actually, most of the living organisms, past and present, have been and are being guided along by the occupants of this invisible world. The consumption of fermented foods has played a very important role in shaping the very intimate relationship we have with microbes. 'Out of sight – out of mind' was the operative dictum, until these obscure creatures became visible, an event that only occurred very recently in human history. In 1635, Van Leeuwenhoek, a Dutch naturalist, published the grounding breaking observations on the association of microorganisms in his own teeth. He described these tiny living organisms as 'animalcules' and suddenly, just like with a magical wand, this obscure inanimate world became clear and bustling with life. Our environment was rendered transparent and vivid. The miniscule lives were everywhere, in rain water, pepper-corn and almost in anything he placed under the microscope. Understanding what we can see then became a reality. However, the microbial world did not seem to be an important territory to explore, perhaps the search of new land far away had grabbed the attention of the deciders of the time. This was the golden epoch for flying the national flag on any land that could be found, conquered and grabbed. Such a preoccupation may have delayed the conquest of the invisible world, despite the clear signs of abundant life that had been unveiled by the introduction of the amazing microscope. Nearly 150 years elapsed before the interest of society at large started to crystallize around the concept of penetrating the microbial life and understanding its role in our lives (Fig. 1.9).

It was only in 1877 that the French scientist Louis Pasteur demonstrated how fermentation was initiated by living organisms living in our midst. Pure microbial species were isolated and cultured in laboratories. The widely held belief that lower life forms could emerge de novo from decomposing matter was forcefully quashed. This new knowledge generated a sea change of attitudes and instilled the

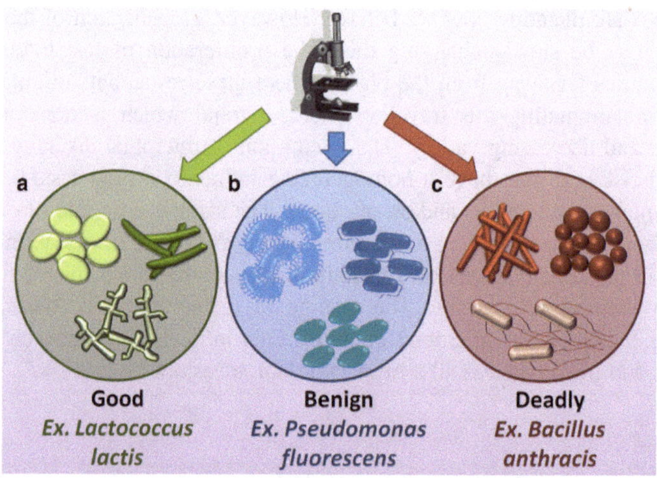

Fig. 1.9 The visualization of microbes (the good, the benign and the ugly). The discovery of the microscope in 1635 allowed to see the invisible world made up of microbes. a Good (*Lactococcus lactis* in dairy products). b The Benign (*Pseudomonas fluorescens*-free living in the soil). c Ugly (*Bacillus anthracis*-lung disease)

notion that we were sharing the planet with the visible and invisible residents. The visible world was made up of all the animals and plants that were on land, coupled with all the creatures in the ocean, which we could see with the naked eyes. The invisible world on the other hand, consisted primarily of minute organisms that were part of the environment and exerted transformative forces on all of us, a fact that was made evident only through the power of the microscope. It became clear that we had developed in association with this invisible universe, however, it was only then being made visible and tangible. Diseases that were the cause of an alarming number of deaths were not due to divine intervention, but were provoked by microbes. Thus, the rush to study microbial life began in earnest in an effort to identify the germs that were behind the misery of so many people and the countless sorrowful events, like the plagues gripping various countries around the globe.

1.8 The Era of 'a Good Microbe is a Dead Microbe'

The discovery by Pasteur of the infectious rod-shaped bacillus that was responsible for anthrax, a deadly disease common until nineteenth century, led to the dawn of the germ theory. Microbes, too small to be seen with the naked eye and visible only through the microscope that invaded human bodies, animals and other living organisms, were the causes of deadly diseases. Microbes became the bane of human existence and the suffering that has befallen humankind. All deadly diseases were the evil work of this microbial population in our midst. Thus, huge sums of monies were invested into studying these microorganisms and to eradicate them. Governments, royalties and financiers alike spent large fortunes to conquer these tiny bugs. This frenetic search led to the discovery of a plethora of disease-causing microbes and to the development of strategies aimed at halting their spread. The main microbial culprits behind illnesses such as small pox, polio, tuberculosis and cholera were unraveled. The ability to isolate and culture these microbes individually provided specific molecular properties of these tiny creatures that were helpful in designing tailor-made remedies. Pasteurization, vaccination, the use of antibiotics and hygiene became the primary tools in combatting these ailments. Immunization, sterilization of milk products and the dissemination of the virtues of personal hygiene contributed to a sharp reduction in infectious diseases. These strategies are still widely exploited to prevent the spread of microbial diseases. The notion that microbes are mostly germs permeates the human psyche even today and 'cleanliness is next to godliness' resonates aloud across the globe. This obsession to rid our surroundings and our living quarters of microbes is imbued in our teachings and quite often results in the over-sanitation of our environment. We are beginning to realize that the overuse of sanitary products may be detrimental to our own good bugs. It is just like utilizing antibiotics to eliminate microbial infections, an act that is giving rise to more resistant and dangerous microbes world-wide. Hence, the jury is still out on the downside of purging our invisible environment constantly (Box 1.4).

Box 1.4 Discovery of Microbes: The Good, the Benign and the Deadly
Following the invention of the microscope in 1635, our invisible world suddenly became a frontier to be explored. The deadly bacteria causing anthrax (*Bacillus anthracis*), leprosy (*Mycobacteria leprae*) and the bubonic plague (*Yersina pestis*) were identified in the late nineteenth century. However, these disease- spawning bacteria were succeeded by the identification of a range of microorganisms involved in a variety of tasks that are critical to the ecosystem and to the well-being of numerous multi-cellular organisms. Free-living microbes (*Azotobacter vinelandii*) that can fix nitrogen and use carbon dioxide as a source of food (*Thiobacillus denificans*) were discovered. The isolation of vitamin B12 from *Streptomyces griseus* in 1948 paved the way for the realization that microbes are essential for the well-being of humans as this essential micronutrient is only produced by the invisible residents of our planet. Subsequently, the contribution of our microbial partners in producing a plethora of valuable goods, like citric acid, amino-acids and pharmaceuticals, was brought to light. Our current day societies would not have existed if is not for the good and benign bacteria, a group that constitutes the bulk of the bacterial world.

From the nineteenth to mid-twentieth centuries the ruling class spent massive amounts of money in the fight against microbial life, as it was seen as an invisible threat to the existence of humanity. Even today, microbes evoke fear in many people. It is quite likely that this war to eliminate disease-causing microbes may have caused some collateral damage on the majority of good microbes that reside on this planet. This may have inadvertently affected the relationship with our microbes and hence our evolution. Indeed, as we will later see, societies that have incorporated the least amount of antibiotics in their medical treatment have more good bacteria in their citizenry when compared to societies that are antibiotic-centric. Low antibiotic societies also have better health outcomes. A microbe-free environment is not conducive to any form of life. It would be life in a bubble. Higher organisms that are devoid of microbes are non-existent, as all multicellular organisms need their invisible partners to accomplish various essential tasks that keep the business of living ticking.

1.9 Transition from Germs to Friendly Microbes

One huge benefit of this public interest in germ theory was the major impact that it has had on the development of technologies to study microbial systems. Scientists started to examine other aspects of microbial existence, particularly how these tiny organisms are essential to the existence of all life forms. The germ-phobic view started to dissipate with the realization that microbes were performing other critical tasks that were crucial to humans. They were the decomposers of waste and the recyclers of the planet. They were literally found everywhere and their purpose was just waiting to be unmasked. The ability to visualize microbial life coupled with the technology to individually grow microbes in the laboratory led to the establishment of the role of microbes in oxygen, carbon and nitrogen cycle in the ecosystem. They were found to be behind many processes that people had taken for granted. They were the ultimate garbage disposers, while at the same time they were also performing the incredible job of being a pivotal contributor to the global protein budget. They execute these undertakings either as free-living organisms or in association with other partners. Indeed, the current atmosphere we enjoy has to a large extent been the work of cyanobacteria, microbial organisms that assiduously fix carbon dioxide into starch and liberate oxygen. The oxygen in the air is essential for all aerobic organisms, including humans, who call this planet home. Microbes are the quintessential decomposers, as they convert all biomass into gaseous and mineral products. Just imagine what will happen to a dead animal or to a dead tree without the intervention of microbial life. The planet would be full of animal and plant debris that would cause a global nightmare. The landscape would be littered with animal and plant remains. It would be a worst sight than the one we have to contend with when our local garbage collection is interrupted for whatever reason. Following the initial stages of decomposition, the dead organisms are converted into minerals, carbon dioxide and ammonia by bacteria. These elements are captured by other microbes and processed into disparate biomasses that triggers the proliferation of other organisms and the cycle of life begins anew (Box 1.5, Fig. 1.10).

Box 1.5 Microbes at the Service of Humankind

Microbes perform unique tasks that allow the ecosystem to function the way it does. They are the ultimate recyclers, converting all living organisms into their elemental components ready to be put back into life. The fate of a fallen tree and a dead animal is the same as they are left to decompose. The final step essentially involves their conversion into gaseous products, such as carbon dioxide and methane and into minerals. This carbon cycle is critical for our survival. It is also an important contributor to the fossil fuels that we so avidly consume. These events have taken eons to form following the decaying of biomass submerged in the oceans and in deserts. Another microbe-dependent process, the nitrogen-cycle, is central to our supply of protein-rich food. The relatively abundant nitrogen in the air is inert and can be processed into ammonia, a precursor to the formation of amino-acids and proteins. The deposition of metals, like gold and copper, is also aided by microbial activities. Whether it is independently or in association with other organisms, microbes participate in a multitude of duties that keep planet earth alive and healthy.

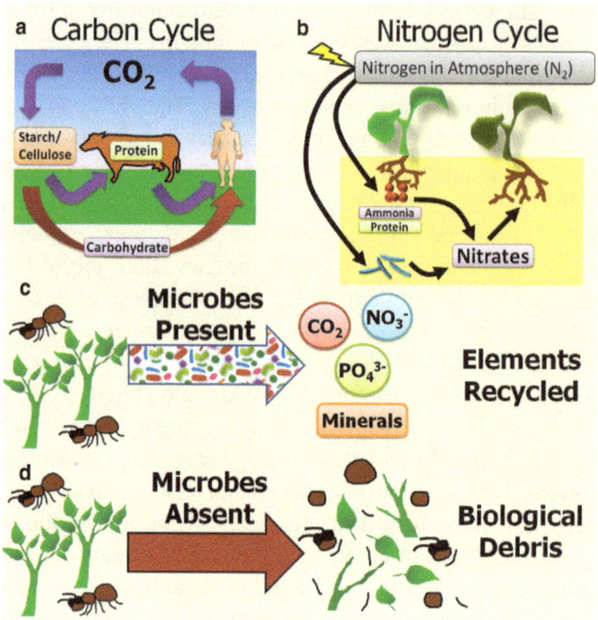

Fig. 1.10 Microbes at the service of humankind. Microbes are involved in a variety of tasks that are essential for the well-being of the planet. a Carbon cycle. b Nitrogen cycle. c Waste management. d A world devoid of microbes will be full of biological debris

Thus, the era of 'a good microbe is a dead microbe' began to shift as the significance of microbial life was unearthed with the introduction of new technologies. These tools allowed us to peer into the invisible life surrounding us. One could better understand what one could see and the door was left wide-open to appreciate how microbes were interacting with other organisms. It became obvious that virtually all life forms had co-evolved with the invisible microbial world. Microbes were part of all organisms making a living on this planet. Many of these creatures would not have been the way we know them today if it was not for the presence of microbes within and on them. The initial concept that microbes were all germs and were the instigators of the deadly plagues ravaging crops and decimating the human population was slowly giving way to a view that good microbes existed which were indispensable to our well-being and our planet. Thus began the era of deciphering how microbial life was contributing to our welfare and those of all living organisms sharing our ecosystem.

1.10 Microbes Within: The Indispensable Invisible Partners in Action

1.10.1 The Nitrogen-Fixers

The discovery in the late nineteen century of bacteria-bearing nodules in the roots of leguminous plants was a turning point in our appreciation of how miniscule organisms invisible to the naked eyes were indeed part and parcel of almost everything alive. Microbial life and multi-cellular organisms are inextricably linked. This vivid demonstration of plant roots harbouring microbes started to quell the widespread anti-microbe feeling, an attitude that had been cemented over many decades in response to the devastation caused by infectious bacteria. Such a view still lingers in some quarters up to today and percolates in the psyche of many individuals. This eye-catching illustration of the visible and the invisible world forging a partnership with each other in order to better their daily lives was a turning point in our understanding of living organisms. The nodules where the microbes were housed were clearly morphologically evident, either in the roots or the stems of plants. How many of us have been amazed to see those miniscule root swellings laden with bacteria in bean plants in our gardens during the summer months. These tiny tireless workers should be praised whenever we are enjoying the protein-rich beans. Plants get precious nitrogen nutrients while the bacteria have a safe place to stay and ample food to partake in. These plants are able to produce beans and peas rich in proteins only due to the presence of the bacteria-filled nodules in their roots. The invisible workhorses laboriously convert nitrogen, an inert commodity, into ammonia, a product which then can be assembled into a range of delectable proteins, which in turn also allows the plants to get their fix of nitrogen. The task of biologically transforming nitrogen has been bestowed only on few organisms and microbes are one of them. Without this natural arrangement, legume-producing plants would need a huge dose of fertilizers loaded with nitrogen chemicals. In fact, an enormous amounts of energy needs to be expended if we are to chemically transform nitrogen, an inert gas that is plentiful in the air, into ammonia, the nutrient plants need to grow well. We must of course be thankful to the tiny nitrogen manipulators each time we enjoy soybeans, peanuts and the wide array of leguminous delicacies (Fig. 1.11).

1.11 The Grass Grazers and the Microbial Connection

Actually, we are now beginning to realize that the symbiotic relationship between invisible microbes and visible organisms is relatively widespread in nature. This synergy that was not apparent before, owing to our inability to visualize these processes, has been laid bare due to the advent of novel technologies. These precise molecular tools allow us to interrogate how microbes live and work in harmony with other living organisms. The milk we drink is derived from this microbial ingenuity. Ruminants like cows are herbivores and primarily eat grass to perform their daily routine. However, they are unable to digest most leafy-green plant materials. The complex carbohydrate, cellulose, which is found in leaves, is chopped into simple sugars, like glucose, by cellulase. Cellulase is a special molecular scissors that is designed to mince cellulose into glucose, a nutrient that the cows subsequently manipulate to produce milk. We

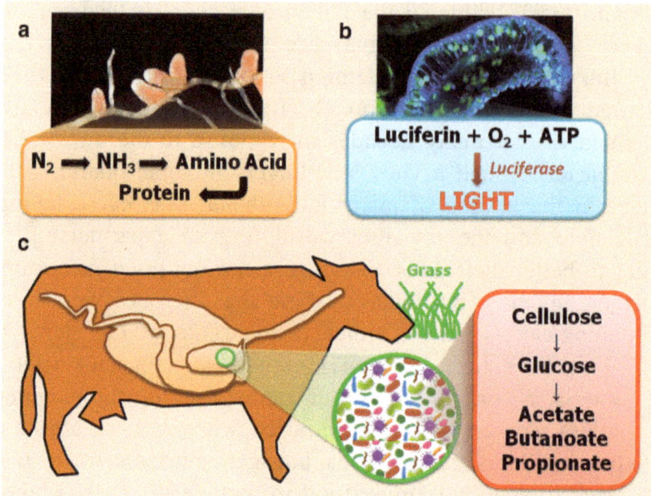

Fig. 1.11 The cooperative and symbiotic nature of life. Microbes in symbiotic partnership executing a variety of life-depending tasks. a Fixing nitrogen (Rhizobium-Legume). b Providing hunting tool (Light-emitting microbes-Squid). c Helping in digestion (cellulose-degrading microbes-Cow)

possess a cousin of cellulase known as amylase that is localized in our saliva and pancreas. Amylase has been instructed to only dice starch, the complex carbohydrate in pasta, potatoes or rice, into glucose. The minute structural variation in these two glucose containing foods necessitates two different enzymatic clippers. Only a few organisms which have made this planet their home have the cellulose-degrading enzyme. The world could readily solve the hunger problem if this enzyme was present in humans.

Microbes may again come to our rescue someday. In fact, all herbivores have an abundance of these cellulose-cutting microbes and most probably have evolved in harmony with them. The paradox is that cows, most herbivores for that matter, consume a staple food that they cannot digest. This conundrum is solved by the embrace of microbial life. Microbes do possess the cellulose-clipper, cellulase and other enzymes that allow cows thrive on a vegetarian diet. Hence, a perfect camaraderie has been struck, leading to some fascinating anatomical outcomes. During their evolution, cows have developed an extra stomach that is in fact the home of the huge amounts of bacteria that help in the production of nutrients that are easily processed by the host. In this biological entente, the bacteria have a safe abode where nutrients are delivered at their doorstep while the cow gets a nano-factory, which is well-oiled to transform grass into goods the four-legged mammal can utilize to grow and produce milk that many of us cannot do without. Next time you drink milk, think of the hard-work of the invisible cellulose-choppers (Box 1.6). Without them our daily dose of delicious dairy products would not be possible. Hooray to the microbes!!!

Box 1.6 Milk Production: A Microbial Connection
The safe environment that the ruminants provide for the microbes enables these invisible partners to chop the grass (cellulose-rich food) into glucose and simple nutrients that are readily utilized by the hosts. Even though herbivores live on a staple nutrient laden with cellulose, they are actually unable to digest this food. Cellulase, the specific scissors designed to clip cellulose into glucose, has been bestowed on the microbes living within ruminants, like cows. Once cellulose is chopped into glucose and broken down into simple nutrients, these are readily utilized by the cow to live on and to produce milk. Hence this mutually beneficial partnership that can be summarized as:

$$\underset{(\text{cellulose})}{\text{grass}} \xrightarrow[(\text{microbes})]{\text{cellulase}} \underset{(\text{cow})}{\text{glucose}} \rightarrow \text{milk} \rightarrow \text{dairy products}$$

1.12 Squids: Microbial Flash Light and Sentinel

Modern molecular tools are permitting us to realize that most, if not all organisms are more than meets the eye. In fact, they are supra- or meta-organisms, which are composed of huge numbers of invisible creatures that have partnered in a common effort to fulfill their biological existence and lead a productive life. This observation is nowhere more evident than in the harmonious pact forged between some squids and their resident bacteria. The microbes inhabiting these squids provide light at night, a key feature in their hunting strategy. The bright illumination emitted by the microbes in the night sky provides a camouflage for the squids' shadow and enables them to trick their prey into the realization that there is no predator around. Of course, the prey becomes an easy meal. For this hunting task, to which the microbes have been assigned, they are rewarded with a healthy diet of sugar and amino acids and a safe squid home to live in. Without the assistance of microbes in shining the bright light at night, the squids would have been left lurking literally in the dark for their meals. The microbes do not only brighten the life of some organisms, they can also act as watchman. In this case, bacteria are recruited to act as sentinels, ensuring that no intruders snatch their prized possessions. To ensure the survival and propagation of their species, squids lay eggs together with a protective layer of bacteria, whose primary occupation is to see to it that these delicious goodies do not become meals for the plundering predators in their midst. The microbes secrete chemical weapons aimed at neutralizing the assault mounted on these eggs by foraging fungi. This stratagem safeguards the hatching of the baby squid and the continuation of the life cycle of this marine organism (Box 1.7).

Box 1.7 Microbial Helping Hand Assisting other Organisms
Bacterial light can be utilized as a hunting tool and is part of the arsenal that many marine organisms resort to acquire their daily food. The light organ is specially designed to shelter the bacteria bestowed with the light-emitting compound known as luciferin. With the aid of the enzyme luciferase, a fatty acid and the reducing factor flavin mononucleotide, light is generated at will to startle and dupe the prey to become an easy meal for the host. The gutless worm has no mouth, digestive tract or excretory system thanks to the microbes it harbours. The microbes convert carbon dioxide, hydrogen sulphide, sulphate and acetate into nutrients for the host. Any waste, like urea or ammonia, produced by the worm is converted into food by the microbes, thus rendering their elimination obsolete. Bacteria residing in numerous organisms are known to help in acquiring essential nutrients. For instance, insects are generally not equipped to reduce oxidized sulphur. However, they need sulphur containing amino-acids like cysteine to survive. This task is left to the bacteria within that reduce sulphate to sulphide for its incorporation in a variety of essential biomolecules. Bacteria also possess unique enzymes like chitinases that help the digestion of krills in whales. Microbial ligninases participate in the metabolism of lignin in insects. Lignins are found in plants and only few organisms can digest these complex molecules. These examples are vivid testimony to the pivotal role microbes play in the development of higher organisms. Most if not all multicellular organisms are the way they are morphologically and physiologically because of the microbes within and on them.

1.13 The Gutless Worm: They Are the Way They Are Because of Their Microbial Partners

The close-associations with microbes assist in easing the daily life of a variety of organisms by providing nutrients, by helping in their hunting ventures and by protecting the survivability of their lineages. However, in some cases the contribution of microbes has gone so far as to obviate the need of some specific organs in living systems – they have literally substituted microbes for a bonafide organ. This microbial presence within has enabled these organisms to have unique morphological features as numerous functions are fulfilled by the resident microbes. In a way, organisms have evolved not to have a specific organ as the job has been taken over by the microbial partners. Hence, the development of the organism has been orchestrated to some extent by microbial input. As these microbes were fulfilling this function, the organ has been made redundant and was eliminated from its anatomical features that are present in their counterparts devoid of these select microbes. This occurrence clearly points to

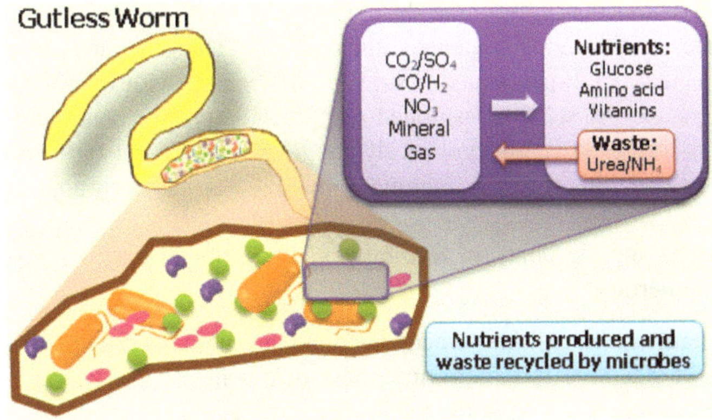

Fig. 1.12 Microbes in lieu of regular organs. Microbes preclude the need of bona fide physiological organs. A Gutless worm (without food intake or excretory organs)

the role microbes have played in shaping and molding the morphological features of multi-cellular organisms. Because of the microbes they harbour some organisms do not have some specific organs as the resident bacteria perform the tasks that would have otherwise been executed by these organs. The association of microbes with worms in the hydrothermal vent has precluded the need of a digestive and excretion system. In this instance, the microbes have rendered the complete food intake, processing and waste removal machinery redundant. In these gutless worms, located in the depth of the oceans, in-house microbes supply them with readily absorbable nutrients. The bacteria fabricate the nutrients from simple minerals that they process into products that are metabolized without the need for a digestive and waste disposal organ, a daily ritual all human beings are keenly aware of. The existence of such morphological features would be impossible without microbial input. Hence, it is reasonable to infer that the evolution of these organisms if not most living systems has been chaperoned by the ubiquitous microbes in their midst (Fig. 1.12).

1.14 The Human and Microbe Bond: How Microbes Became an Integral Part of Us

These few examples help illustrate how symbiosis involving microbes and multicellular organisms is beneficial to all the participants. This is a more widespread phenomenon in nature than previously thought. These mutually profitable arrangements are the hallmarks of virtually all living systems and enable the proper functioning of these supra-organisms. It is then sensible to conclude that life would not have existed in its current form without the presence of microbes and the inherent cues they send. These signals subsequently shape the different life forms.

Microbes play a critical role in the development of both physiological and anatomical features of the living organisms they inhabit. Thus, it stands to reason that humans too must have negotiated a similar relationship with microbes and these invisible creatures must have fashioned our physiology, our being and follow us in all of our living processes and in all our tribulations from birth to death and beyond. However, this microbe – human bonding has not been appreciated until recently.

The reasons for the delay in recognizing and embracing this invisible organ have been numerous:

1. In our germ-centric culture, where microbes are perceived as the harbinger of diseases, it was hard to fathom how these tiny beings could be part and parcel of ourselves and assist us in the business of living.
2. As microbes are invisible, it is hard to comprehend whatever we cannot see – this also applied to our initial understanding of the earth as being flat until we were able to appreciate its globular nature by circumnavigating the earth.
3. The anthropocentric nature of our culture, where humans are the superior beings and above the rest of living organisms, has shaped us not to include any other living systems as an integral part of ourselves; this affirmation would make us to be a mere component of the complex natural landscape that human cultures have sought to be master since time immemorial.
4. And lastly, of course, this intimate microbial and human pact will go counter to many ingrained religious beliefs, where the invisible world is barely ever the part of human history.

The large creatures that were in full view to the naked eyes were obviously front and center – one cannot comprehend what one cannot see nor is aware of. That was and has been the case with microbes until recently. These dogmas are fast disappearing as the unveiling of this microbial life has been made relatively easy, tangible, lively and effective with the discoveries of new technologies. Microbes can be visualized within their communities in our body. The chemical fingerprints they generate, mirror the critical functions they perform for their human hosts. For instance, the production of vitamin K, a key component of the blood coagulation process, can be readily tracked to our microbial partners in the gut; we cannot produce this coagulation factor and would be deprived of the clotting property of our blood.

1.15 Visualization of Microbes Within Humans: The New Era

The current medical and microbiological education is built on infectious diseases and how humans respond to these invisible invaders. It will not be too long before this concept will be radically transformed to view microbes residing within and on us as a human organ fulfilling a variety of critical tasks to keep our body functioning properly. We will never survive our natural habitat without microbes dwelling in and on our bodies. If we are to be microbe-free, we will have to live in a bubble – which is not an enticing prospect. The Human Microbiome Project (HMP), initiated in 2007 to map the microbes humans possess, is revealing fascinating and intriguing details on our association with microbes. Indeed, it is becoming abundantly clear that we are the way we are because of the microbes dwelling within and on us. This initiative is bound to unleash a treasure trove of information, confirming the pivotal role microbes play in our health and wellness. These findings are going to be akin to the discovery of the invisible world following the invention of the microscope. This magnifying tool lay bare to our eyes our invisible surrounding; we suddenly realized we are living among tiny living organisms that are ubiquitous and accompany us everywhere.

With the current endeavor, the veil will be removed on the nature of the microbes we live with and the functions they perform for us within and on our body. Although the realization of the omnipresence of microbes within us is becoming more transparent, the first observation that humans have intimate fruitful interactions with bacteria was made in 1885 by the Austrian pediatrician, T. Escherich. He reported the presence of colon bacillus currently known as *Escherichia coli* in the fecal matter of healthy children and those suffering from diarrhea. Today, we know that 40–60% of stool is made up of microbes, among which *E. coli*, a bacterium named after Escherich, is a major component. Following this first recorded observation of the link of bacteria with the human body, several documents on the presence of microbial flora in the mouth, skin, digestive, respiratory and urinary tracts were reported. Although there was an appreciation that these microorganisms may be contributing to the general well-being of the host, they were mostly viewed with suspicion and as disease-causing. This notion was prevalent, despite the fact that bacteria supply the human body with vitamins B and K, ingredients essential for our existence. In fact, vitamin B12, a vital micronutrient for most animals, including humans, can only be produced by microbes. A deficiency in this vitamin gives rise to a medical condition known as pernicious anemia. It also affects the body's ability to metabolize fatty acids and can give rise to numerous other abnormalities. Why is an ingredient so critical for our survival is only available from the microbes residing in our gut or from foods (meat and poultry) we eat, where it is also produced by microbial intervention? It is not even synthesized by plants. Why would nature make such an arrangement if it was not for the pivotal role of microbes in our development? But the war mentality against germs permeated society so much that isolating deadly microbes was the most urgent task that the microbiological and medical communities were engaged in. In this frenzy to find the deadly bugs that were responsible for the numerous plagues in the twentieth century, the contribution of microbes to human health was largely ignored (Box 1.8).

Box 1.8 Vitamin B12 – The Link Between Microbes and Humans
Vitamin B12, also known as cobalamin, contains cobalt and is an essential nutrient in humans. However, it can only be synthesized by microbes, even plants cannot produce this ingredient. Vitamin B12 was isolated in 1948 and its role in triggering pernicious anemia, an auto-immune disease leading to gastric atrophy, was subsequently established. It is involved in lipid, amino acid, nucleic acid and fatty acid metabolism. Its deficiency has been associated with cognitive impairment, vascular disease and immature red blood cells. Increased levels of homocysteine are a hallmark of vitamin B12 deficiency. The recommended dietary allowance (RDA) for adults is 2.4 ug/day. The following products are a good source of vitamin B12: tuna, sardines, salmon, meat, eggs, milk and cheese. Strict vegetarians need to be nutritionally vigilant as vitamin B12 is not present in plants.

It was not until very recently that the concept of 'a good microbe is a dead microbe' started to give way to the idea that humans, like most other living organisms, harbour an enormous amount of bacteria. This is becoming widely accepted as new techniques are providing a clearer picture of the tasks that these microbes are performing for the human body. We are indeed now beginning to understand the real us, as novel technologies are exposing the activities microbes are executing within and on our bodies. Intestinal microbes are immensely beneficial to human physiology and metabolism as they perform a myriad of indispensable duties that ensure our well-being. They digest complex carbohydrates; they contribute to the pool of essential amino acids, like leucine, and stimulate the production of fats. Numerous studies have revealed that germ-free (gnotobiotic) animals cannot survive starvation as long as their normal microbe containing counterparts can. Despite the unequivocal link between the human body and microbes, the technology was not able to capture the microbial communities residing at the location where they were performing their jobs – that was until the discovery of the polymerase chain reaction (PCR) and next generation sequencing (NGS). The addition of these two techniques have given us a unique opportunity to visualize the invisible organ that has been with us forever, emitting signals but never seen and understood. It was the ingenious aggrandizement of the power of PCR that led to the block-bluster movie Jurassic Park. The DNA of dinosaurs that was preserved in the fossils of insects that had fed on them was amplified and reconstituted into a living animal and beyond. This technology offers the tremendous advantage of working with infinitesimal amounts of nuclear material, a real boon where organisms cannot be cultured or regrown.

In fact, only a minute fraction of microbes living in nature can be isolated and cultured. Although the vast amount of microbes thrives happily in their respective communities, they cannot be grown individually, meaning that they cannot live alone, they need their communal lifestyle to survive and proliferate. They might have an absolute requirement for certain ingredients produced in the community

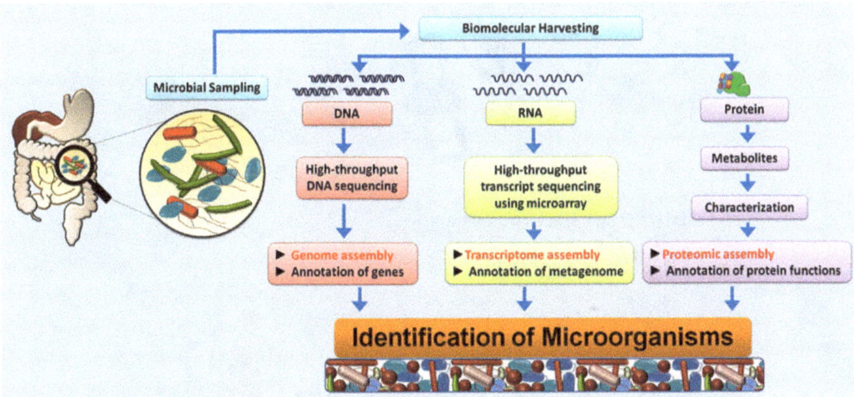

Fig. 1.13 Molecular imaging technologies. These culture-independent technologies have the capability of identifying the microbes and also the activities they are performing. DNA and RNA are indicative of the uniqueness of each microbe; Proteins and metabolites provide a window on the task they performing in the community

to thrive. These harmonious microbial ecosystems have taken a long time to evolve and are almost impossible to recreate in a laboratory setting. This dilemma has been resolved with the fine-tuning of molecular technologies like PCR, NGS, 16S ribosomal RNA (rRNA) fingerprinting and DNA profiling (Fig. 1.13). These procedures allow the visualization of this invisible microbial world, help with the identification of a specific bacterium and can lay bare the secrets of their biological prowess. These tools are revealing this new microbial world in the same way the microscope helped uncover our invisible environment in the seventeenth century. Now we can interrogate and communicate with virtually all these microbial residents without missing any of them. Suddenly, we can see and explain biological processes that we knew were present, but were unable to visualize. The PCR is a sheer magical procedure that lets us make a huge number of copies from a single piece of DNA, while NGS provides a unique means of studying microbes without even culturing them; they can be picked right from their natural communities and identified. The minute quantities of genetic material that are isolated from these natural habitats can be amplified and their lineages detected through various bioinformatic applications. It is more or less like the global positioning system (GPS) that can identify any individual anywhere on the planet (Fig. 1.14).

Furthermore, the proteins and the metabolites that are the workhorses of all organisms can also provide us pertinent information on the activities these microbial populations are performing. These technologies allow us to catch them right in action while they are executing. The goings and comings of the invisible community are exposed. It is a bit like identifying you say from your spit and also detecting what you just ate or from your sweat and at the same time pinpointing the activity you were engaged in. In this manner, the profile, the fingerprint and the activity being conducted at a given moment by each microbial community can

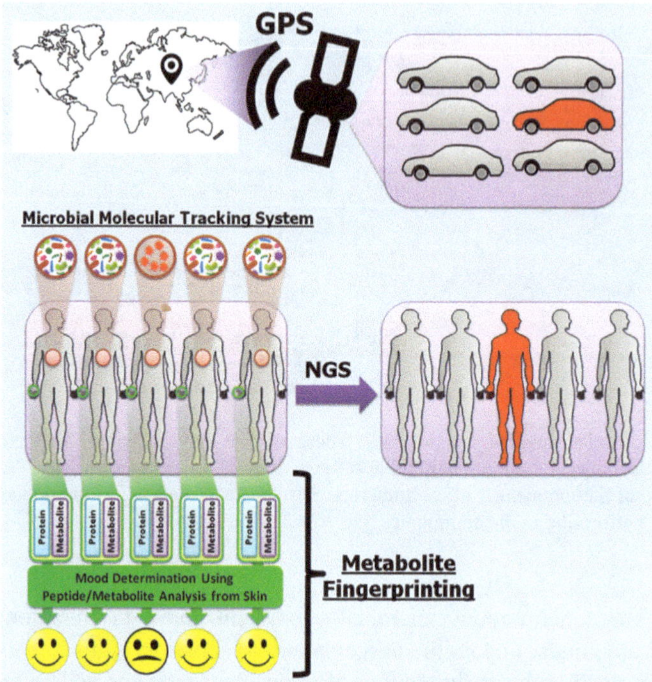

Fig. 1.14 Molecular imaging of microbes where they reside. The culture-independent techniques allow to identify the microbes through DNA/RNA analyses and to assign their functional characteristics by monitoring metabolites and proteins. The detection of a uniquely tagged vehicle by a global positioning system (GPS) and the identification of an individual by tracking the resident microbes with NGS

be tracked and documented. The participation and the role of each community member can be identified. These tools have rendered the intimate interactions between our resident microbes and our own body relatively easy to follow and reveal why we are a supra-being with 10^{-27} cells, the majority of which are microbial in origin. The application of this technology that is effective, inexpensive, and relatively easily accessible has made the task of cataloguing our invisible self within reach. Once, the categorizing of the microbes and the territory where they reside in has been undertaken then other attributes of this invisible organ may become more within our grasp. It will be akin to when we recently saw the first visual of the blue planet charting its course around the sun from the space station. Even though we are all aware that this phenomenon occurs, this is the first time in 2016 we had a slow motion view of this event. Microbes are part of us, we know they perform enormous tasks within our body, we will now be able to appreciate 'in person' how they are working for us just in the same way we have seen the heart pumping blood through our arteries. These fascinating molecular imaging techniques will confirm the existence of the physiologically active microbial communities within and on us that have not been fully recognized hitherto (Box 1.9).

Box 1.9 New Technologies and the Molecular Imaging of Microbes
Organisms can be identified and their physiological attributes deciphered by profiling the information they harbour (DNA, RNA) and functional chemicals (proteins, metabolites) they produce. These entities even in minute quantities can be picked right in situ, analyzed and functional properties established. These culture-independent techniques can now track the vast amounts of microbes that were hitherto impossible to visualize. The instruction that specifies and dictates an organism resides in the DNA and a set of DNA that relays the commands governing any characteristic or function is referred to as the gene(s). These are located in the chromosomes where the information is securely kept. The advent of the PCR, a technique that can amplify a single copy of DNA (gene) into several thousand copies has afforded the ability to identify even the sparsest of microbial population. The 16S rRNA technology has made it simpler to classify microbes and allocate their relationship to a specific family (phylogeny) they belong to. During the biological information transmission DNA is copied into RNA. There are different types of RNA in an organism, those residing in the ribosome known as rRNA. A variety of these referred as the 16S rRNA is highly conserved and contains the fingerprints that are utilized in the classification and identification the microbes. Following the identification of the microorganisms, the proteins and the metabolites that provide a snapshot of the activities they are engaged in can be profiled. For instance an active enzyme like cytochrome C oxidase will point to aerobic respiration while the presence of metabolit like formate, ethanol or lactate may reveal life in oxygen-limited environments. These molecular techniques not only reveal the communities these microbes are living in but also divulge how they are contributing to the communal lifestyle. This molecular imaging technology has opened new vistas on our invisible partners. The NGS and the shotgun metagenomic sequencing platforms that incorporate many of these analytical techniques have enabled the interrogation of host-associated microbial communities, community diversity and function, features crucial in understanding the interaction between the invisible world and the visible organisms. Information storage (DNA-gene-chromosome-information box; communities); Information transmission (DNA.RNA, Proteins, metabolites – molecular imaging-identification of community members and activities).

1.16 The Human Micribiome Project: Cataloguing Our Invisible Self

The HMP was launched initially with the goal of systematically identifying, and categorizing the microbial population in humans in order to elucidate the role of these organisms play in the body. The collection of all microbes comprising bacteria, viruses, archaea and eukaryotes are referred to as the microbiome. The variation of the microbiome with age, sex, ethnicity, geographical location and social interaction is an important mandate of this project as is the contribution of these microbes to our well-being.

The knowledge generated will allow to piece together the contribution of our microbes to our diversity, specificity and the working of the human body with all its components, visible and invisible. All, if not most living organisms have co-evolved with the first residents of this planet who started their earthly journey 3.6 billion years ago, eons before any creature. Humans and microbes have forged a unique alliance that is beneficial to both participants. Humans will not be the same without their resident microbial friends and it is becoming increasing evident they have a say in our anatomical and physiological condition. This delicate balancing act can be readily perturbed by a variety of factors that are only now beginning to be unraveled. The disruption of this harmony has major implication for our health hence the concerted global effort to uncover the workings of the microbiome will pave the way for innovative practices in the healthcare delivery system and in promoting our well-being. Once the veil on this now invisible organ is lifted, the constituent microbial cells will become an important component of health education. The culture-independent molecular-visualization technology will not only help catalogue each individual microbial composition at strain specific level but will capture the dynamics and functions of the microbial communities. It is like a satellite reconnaissance system from space pinpointing someone in a crowd on earth and zeroing on the activity this individual is engaged in at that particular time. So far the results have been tantalizing (Box 1.10).

Box 1.10 Human Microbiome Project: Goals and Potentials

The HMP was initiated by the National Institute of Health in 2007 with the objective of elucidating the microbes residing within and on the human body. The microbiological data generated, the innovative technologies developed and the computational programs created to synthesize this information are intended to be utilized in an effort to improve human health and the health care delivery system. This knowledge is expected to revolutionize medical education as the microbes that constitute a vast amount of cells and genetic controls will be an important component of who we are physiological and biochemically. It will advance our understanding of various diseases and enable the design of health solutions aimed at specific individual where resident microbes, the invisible organ will be viewed as an integral component as critical as the visible organs like the liver. This initial project has taken an international urgency as numerous countries including Canada, Japan, France, and the European Union have joined in the mapping of the microbes inhabiting humans. Similar projects aimed at cataloguing other animals and plants have sprung globally. The realization that most multi-cellular organisms are indeed supra-organisms with their visible and invisible parts is quickly beginning to be etched in all scientific endeavours aimed at living organisms. By helping establish the origin, the development, the variation and the role of the microbial communities within humans, the HMP will aid to clarify numerous biological enigmas health professionals are confronted daily, an objective that will benefit society at large.

1.17 A Glimpse at the Human Microbiome

Let us explore what we are made up of. A human being is in reality a supra-organism with approximately 10^{13} cells inherited from his or her parents and 10^{14} microbial cells. We all possess many more cells than our own. These are in fact microbial cells with their own genetic fabric and mind. While our own genes amount to about 23,000, the genes of microbial origin can range between 2 to 3 million. They constitute 1–3% of our body mass and in an 80 kg adult the total mass of the microbes may fluctuate from 2 to 3 kg. For comparison, the mass of the average adult human heart is between 0.22 to 0.42 kg, making the microbial size at least five times bigger. There are almost 50 phyla (family) of bacteria. However, the most common six phyla are the *Firmicutes, Bacteriodes, Proteobacteria, Actinobacteria, Fusobacteria and Verucomicrobia* (Fig. 1.15).

Microbes inhabit virtually all the nooks and crannies in and on our body like the nose, the skin, lung or the gut. However, the type and amount of microbes vary significantly. For instance, the gut contains the largest share. The evaluation of the microbial communities in the gut is yielding information on the core and non-core members in the different regions of the intestine. So far there appears to be around 40 microbial subgroups that constitute the partially stable members of the gut microbiome. Even in the same organ, there is a variation in microbial communities. The proximal end of the small intestine favours the *Firmicutes*. This territory is not ripe for colonization by the members of *Bacteriodetes*; the distal region (colon and caecum) on the other hand, is a fertile ground for the *Actinobacteria* and

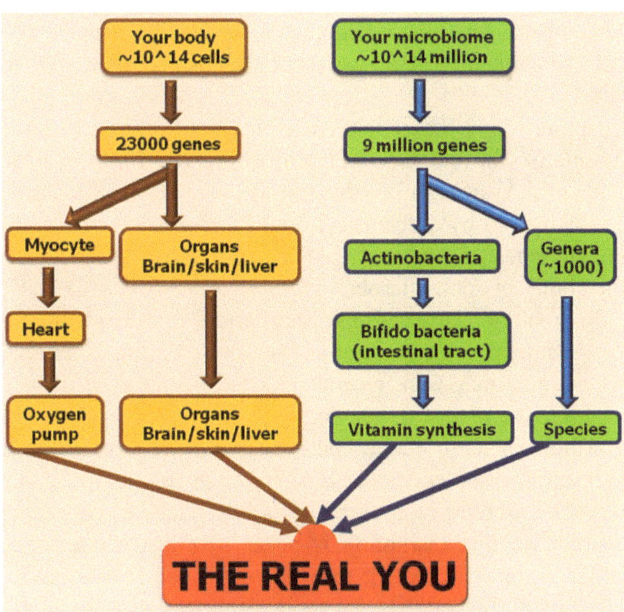

Fig. 1.15 The real human with all cells and genetic make-up

Proteobacteria family members. They are involved in a variety of tasks that make our lives easier and enable us to live in harmony with our environment. These will be explored in following chapters. The very intricate communication between the host and some of the microbial communities is divulging how the host has a major say on the specificity of the microbial population she or he wants to see flourish. The host provides a protected area and select nutrients where only the desired microbes can lay claim to that piece of human real estate. The mucous layer in the colon with the abundance of the complex carbohydrate like mucin only permits the colonization by microbes that can utilize these nutrients. *Bacteriodes fragilis and Akkermansia muciniphilia*, two of the mucin-degraders are privileged in these protected habitats as we will later see how they perform admirably for the host. This finding has an uncanny resemblance to what has been uncovered many decades ago in the relationship between the nitrogen-fixing rhizobia and leguminous plants. To get access to their homes in the plant roots, the rhizobia secrete polysaccharides (complex carbohydrates) that are utilized to unlock the doors to the plant roots. Once this signal is recognized by the plants, they happily open their arms to accept their symbiotic friends. This communication is very host specific that is one legume takes one select nitrogen-fixer while a different plant is adapted to another rhizobium. The mucin in the colon coupled with some other chemical messenger yet to be unearthed may be corralling the proper microbial community by enticing the desired dwellers with food that other micro-organisms cannot utilize. Indeed in hibernating mammals the abundance of these mucin-degraders increases as they contribute to the energy budget of their host by supplying them with ready-made metabolic fuel. The ability of the stomach-inhabitants, the *Helicobacter pylori* to cling to the glycan hooks on the gastric lining enables them to colonize the inhospitable acidic environment they are confronted with. The chemical gradient, the pH, oxygen levels, nutrient concentration and the compartmentalized host immune activity all conjugate to establish the proper microbial communities that we are only now beginning to comprehend (Fig. 1.16 and 1.17).

The laser capture micro-dissection procedure is enabling the profiling of the microbial communities in real time and providing vivid imagery of the dynamic nature of our microbiota in both healthy and diseased states. For instance, hepatic encephalopathy is marred by a decrease in the *Roseburia* species and by an abundance of such microbes as *Enteroccus*, *Veillonella* and *Bifidobacterium*. Recent findings on the ability of such microbes as *Ruminococcus gnavus* and *Clostridium symbiosis* or the prebiotic sialylated oligosaccharides to reverse the growth deficit augur well for the integration of microbial therapy in curing a variety of ailments (Fig. 1.18). These microbes were recently tracked in the stool of well-nourished children or children with no growth deficit in Malawi- a golden stool indeed. How do we get hold of these magic microbial potions? Is microbial perturbation the cause or a result of diseased states? How it is triggered and who is predisposed to it? Can microbial therapy be a cure for various illnesses and can we fertilize our invisible organ so that disruption is never permitted or to rejuvenate it in the same manner we are attempting to coach stem cells to revitalize moribund,

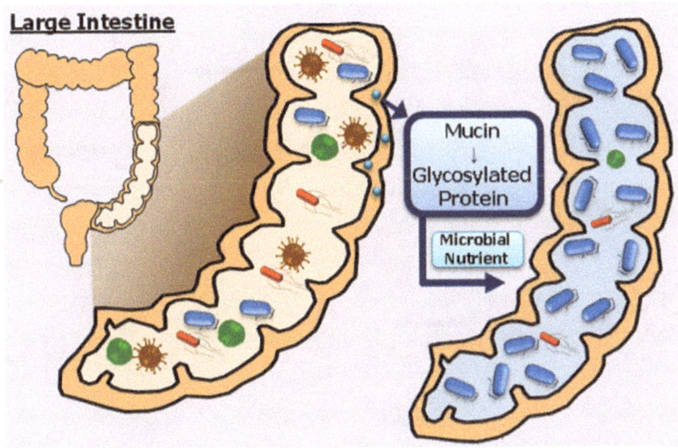

Fig. 1.16 Colonization of the large intestine. Mucin helps attract the proper microbes

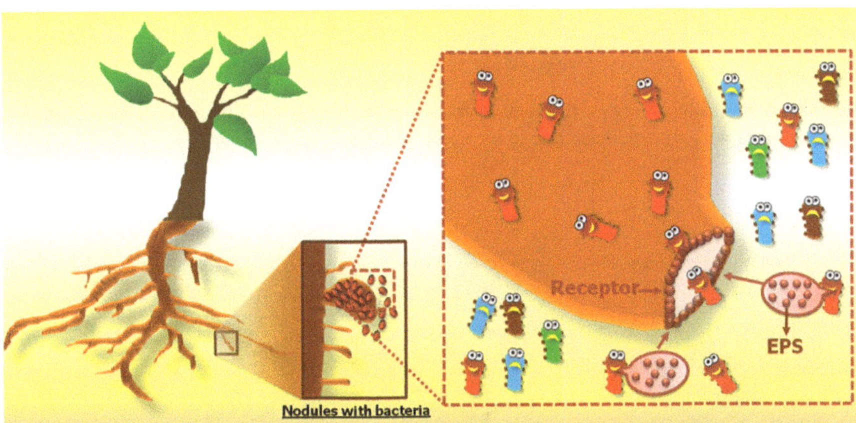

Fig. 1.17 How hosts attract the proper microbial partners. The exopolysaccharides and the metabolites decide on the specific rhizobium that will nodulate a given leguminous host

non-performing or aged visible organs? These are some of the questions that the current focus on the microbiome will help elucidate. We have just skirted the tip of the iceberg. The secrets stored in the invisible organ will revolutionize our relationship with ourselves and with other beings. Our habits, routines and our dealings with all aspects of our lives will be impacted upon. In the following chapters we are going to dissect the ongoing understanding of this 'unseen' organ, its influence on our wellness and how we can contribute to our well-being by taking care of the well-being of our microbiome. We are inextricably linked in our life journey. I will gaze into the microbiome crystal ball in order to piece together what

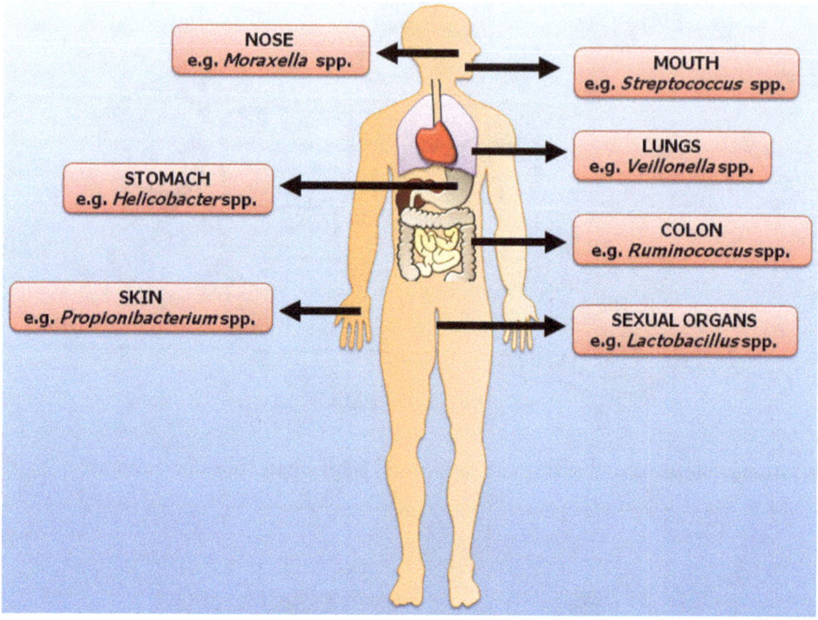

Fig. 1.18 The various body parts where microbes are more prominently embedded. The gut has most microbes while the stomach due to its acidic nature tends to harbour lesser amounts

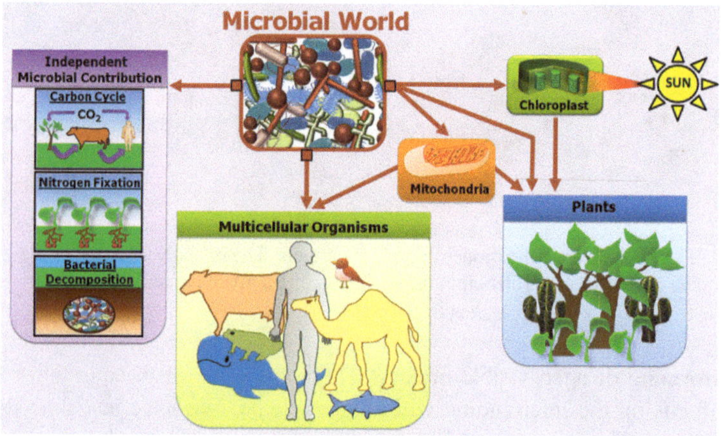

Fig. 1.19 Microbes at the center of life on planet earth: A schematic snapshot

the unraveling of this invisible entity within us has in store for our very own being Let the microbial treasure trove in our midst be unleashed. And salivate on the possibilities these bugs can accomplish for us and society at large as we embark on this thrilling microbial adventure (Fig. 1.19 and 1.20).

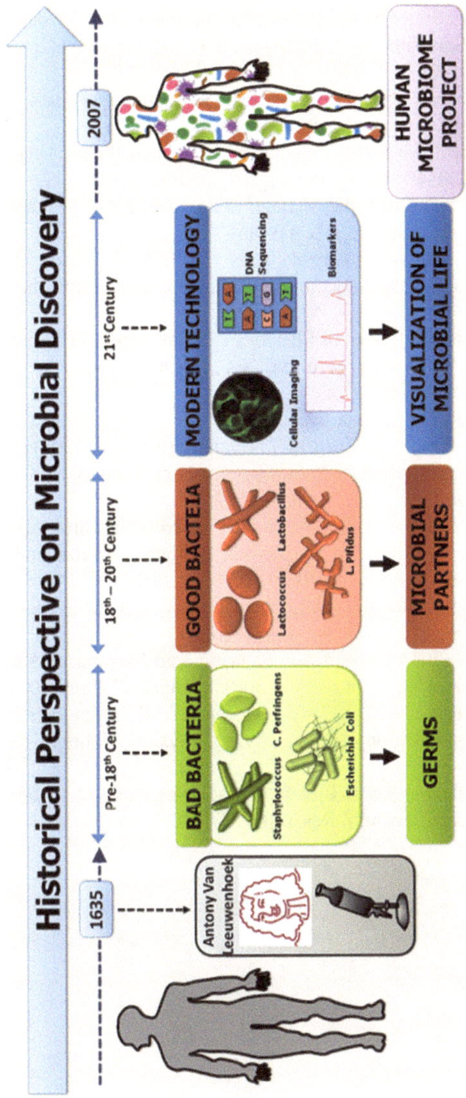

Fig. 1.20 Unveiling the invisible organ within: Historical high-lights

Suggested Readings

Appanna VD, Finn H (1995) Microbial adaptation to iron: a possible role of phosphatidylethanolamine in iron mineral deposition. Biometals 8(2).
Bosch TC, McFall-Ngai MJ (2011) Metaorganisms as the new frontier. Zoology 114(4):185–190. https://doi.org/10.1016/j.zool.2011.04.001
Brachman PS (2003) Infectious diseases–past, present, and future. Int J Epidemiol 32(5):684–686
Carter KC (1985) Ignaz Semmelweis, Carl Mayrhofer, and the rise of germ theory. Med Hist 29(1):33–53
Cavicchioli R (2015) Microbial ecology of Antarctic aquatic systems. Nat Rev Microbiol 13(11):691–706. https://doi.org/10.1038/nrmicro3549
Chilton SN, Burton JP, Reid G (2015) Inclusion of fermented foods in food guides around the world. Nutrients 7(1):390–404. https://doi.org/10.3390/nu7010390
Cho I, Blaser MJ (2012) The human microbiome: at the interface of health and disease. Nat Rev Genet 13(4):260–270. https://doi.org/10.1038/nrg3182
Fonseca-Garcia C, Coleman-Derr D, Garrido E et al (2016) The cacti microbiome: interplay between habitat-filtering and host-specificity. Front Microbiol 7:150. https://doi.org/10.3389/fmicb.2016.00150
Franzosa EA, Hsu T, Sirota-Madi A et al (2015) Sequencing and beyond: integrating molecular 'omics' for microbial community profiling. Nat Rev Microbiol 13(6):360–372. https://doi.org/10.1038/nrmicro3451
Grice EA, Segre JA (2012) The human microbiome: our second genome. Annu Rev Genomics Hum Genet 13:151–170. https://doi.org/10.1146/annurev-genom-090711-163814
Human Microbiome Project C (2012) A framework for human microbiome research. Nature 486 (7402):215–221. https://doi.org/10.1038/nature11209
Integrative HMPRNC (2014) The integrative human microbiome project: dynamic analysis of microbiome-host omics profiles during periods of human health and disease. Cell Host Microbe 16(3):276–289. https://doi.org/10.1016/j.chom.2014.08.014
Ji BN J (2015) New insight into the gut microbiome through metagenomics. Adv Genomics Genet 2015(5):15
Rhee SJ, Lee JE, Lee CH (2011) Importance of lactic acid bacteria in Asian fermented foods. Microb Cell Fact 10(Suppl 1):S5. https://doi.org/10.1186/1475-2859-10-S1-S5
Selhub EM, Logan AC, Bested AC (2014) Fermented foods, microbiota, and mental health: ancient practice meets nutritional psychiatry. J Physiol Anthropol 33:2. https://doi.org/10.1186/1880-6805-33-2
Wen TM, Chen MT (2010) Changes of Protein in Natto (a fermented soybean food) Affected by Fermenting Time. Food Sci Technol Res 16(6):537–542

Chapter 2
The Microbiome: Genesis and Functions

Contents

Abstract As in the case of the visible organs, the microbiome also begins its developmental stages in the womb. Further evolution and maturation of the organ then proceed after birth during childhood, puberty and beyond. An array of disparate factors including genetic, geography, and nutrition converge to shape the microbiome each of us possesses. For instance, individuals born in rural Africa or

V.D. Appanna, *Human Microbes - The Power Within*,
https://doi.org/10.1007/978-981-10-7684-8_2

South America tend to have higher amounts of *Prevotella* than their counterparts in Europe or North America. In a similar fashion the lungs are larger in residents living in high elevation than those dwelling at sea-level. The microbiome not only plays a role in the establishment of the visible organs but it also fends against opportunistic microbial invaders, produces vitamins, enzymes, anti-oxidants, neurotransmitters and participate in a plethora of vital functions. The effectiveness of the digestive tract in extracting the maximal amount of nutrients from the foods we intake is largely dependent on the microbial helpers residing in the alimentary canal. The ability of different populations to consume food products available only in their specific region is attributed to this microbial assistance. The digestion of dietary fibres, the activation of drugs and the elimination of toxins are mediated by the wide range of enzymes that are associated with the constituents of the microbiome. Although the molecular workings of the microbiome is only now beginning to be unravelled, it is becoming abundantly clear that these microbial partners are intimately linked to humans, make them the way they are and impart on each individual the unique biological and physical attributes they exhibit.

Keywords Organogenesis · development · digestion · geography · genetic · diet · vitamins · dopamine

2.1 The Initiation and Development of the Microbiome

The seeds of this organ are sowed from the beginning of our lives in our mothers' wombs. These initial building blocks are there even before the existence of many of our visible body parts. In fact, it will not be surprising if we will soon learn that the microbiome may be shaping and contributing to the development of the other organs since our formative months. These invisible partners accompany us from the early stages of our life. Until very recently, the maturation and the progression of human embryo were thought to be taking place in a relatively sterile environment. The placenta was considered to be devoid of any microbe. However, the introduction of new microbial finger-printing technologies have helped debunked this assumption. Indeed, bacteria have been found in the placenta. Prenatal exposure to microbes is a natural part of in utero nurturing. The development of the fetus may be responding to signals generated by these microbial activities. The fetus is immersed in the amniotic fluid that contains microorganisms as does the umbilical cord. The placenta has been found to harbour *Lactobacillus* while the amniotic fluid and umbilical cord provide refuge to microbes like *Streptococcus and Staphylococcus.*

These microbial species may vary with environmental factors that influence the mother's microbiome. The vertical in utero transmission from mother to the developing embryo is a critical component of pre-natal exposure that contributes to the maturation of the unborn and may help shape the child's anatomical and physiological processes once born. Hence, it is crucial to understand the nature of microbes we are subjected to while we are still developing in the womb if we are to have a proper understanding of how the body really evolves and functions. The discovery of microbial life in the fetal meconium, the first stool of the new-born is revealing indeed, as it provides a vivid evidence of our first encounter with our invisible life-partners. As mentioned before one can even safely suggest that they are with us even before the development of the visible organs and biological systems. The possibility that they play a crucial role in moulding the evolution of our organs and physiological behaviour is becoming more evident. With the technological tools at our disposal and the new techniques being introduced, it will not be too long before the significance of the first encounter with our microbial partners is elucidated.

2.2 Our Invisible Organ: Parents and Location Matter

Following the forging of this friendship before we are born, we start acquiring the other constituents of our invisible organ from the environment we live in (Fig. 2.1). Depending where and which itinerary we take to come to planet

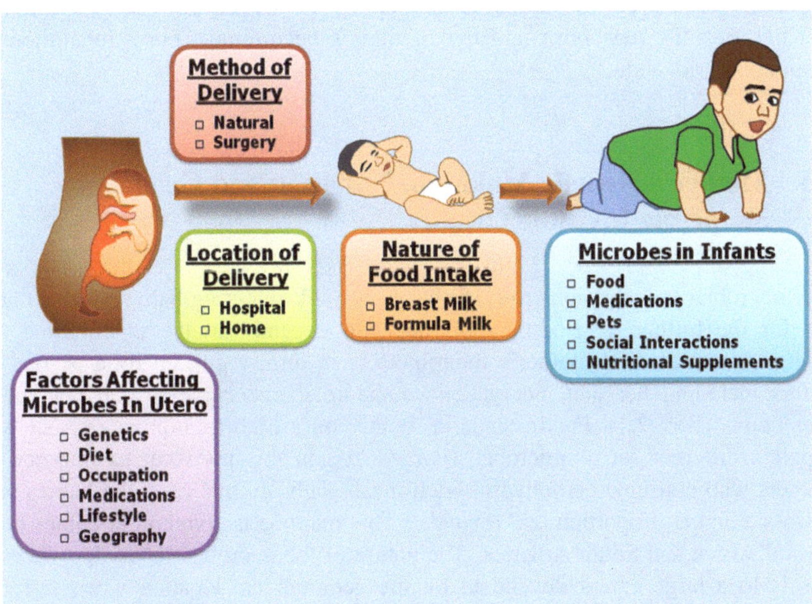

Fig. 2.1 The acquisition of the microbiome. The microbiome we acquire at birth depends on a variety of factors including the genetic make-up of the mother, mother's microbiota, mode/place of delivery, medications, and diet

earth, our microbial flora bears the scars of this journey and that of the location where we are in. The microbial population may vary depending whether we are delivered by caesarean (C-) section or by vaginal route. In babies born by C-section, the mother's skin bacteria and those present in the hospital are prominent. *Staphylococcus* and the dermal *Propionobacterium* seem to dominate while the amount of *Bifidobacteria* is relatively lower. They also harbour higher levels of facultative anaerobes, microbes known for their negative influence on our health. Microbes such as *Lactobacillus* and *Prevotella* are more abundant in those babies born naturally. Pre-term babies also have an entirely different microbial population and the level of *Bifidobacteria* is lower than in full-term babies. Hence, the journey we take to get to earth may dictate the constituents of the invisible organ we possess while we are a child and beyond. This in turn guides our microbial landscape and the workings of our other body parts as we progressed through life. In fact, the microbes we inherit from our mother help us navigate the danger associated with our first infection. Children born through C-section are more prone to asthma and allergies and experience a delay in microbial colonization compared to those delivered without surgical intervention. Recent research is revealing how swabbing C-section babies with the mothers' vaginal microbiota is reversing allergies, asthma, infections and other maladies that these infants have problem fighting. This intervention may become a common practice in hospitals where these babies can get an inoculum of their mothers' microbes as a booster that will arm them to confront the invisible dangers of the world. Skin to skin interaction between the new born and the mother is becoming a norm in numerous hospitals world-wide.

2.3 Mother's Genetic Make-up, Nutrition and Geography Shape the Invisible Organ

This microbial transmission from mother to the still maturing baby that is so crucial for the further development of the child depends on the microbiome the mother possesses. The mother's microbiota is in turn, regulated by a myriad of factors including her diet, her genetic make-up, her occupation and her social interactions (Box 2.1). For instance, a mother on a high fat/high sugar diet will impart a different set of microbes (*Bacteroides* in this instance) as opposed to mothers who consume carbohydrate-rich meals daily. In this case, the babies will possess a higher proportion of *Prevotella*. This microbe is common in babies born in rural Africa and South America. The nature of the microbial landscape we possess is to a large extent dependent on the geographical location where we are born. Northern European babies have an abundance of *Bifidobacteria* while *Lactobacillus* seems to be predominant in those babies seeing the first daylight in Southern Europe.

Box 2.1 Factors Contributing to the Establishment of the Microbiome
The establishment of the microbiome is influenced by a variety of factors. Diet, life-style, location and social structure are some of the modulators of the microbial communities one will have. An infant feeding on the breast milk from a mother who consumes high-fat nutrients will tend to have a microbial population rich in *Enterococcus* and a lower count of *Bacteriodes*. The latter are known to aid in building up the immune system. Infants fed breast milk by mothers on a regular diet will have higher amounts of *Bifidobacteria*, microbes that impart numerous health benefits including an ample supply of vitamins. Nutrients rich in fibre and low in fats as seen in rural Africa and South America promote microbes like *Actinobacteria* while foods with high fat and low fibre contents, a situation common in Western countries provide a fertile environment for *Proteobacteria*. The former is housed in various part of the body including the skin and play an important in fending off opportunistic bacteria. Increased levels of the latter are associated with diseases. The use of antibiotic and medications also has an impact on the microbial ecosystem the infant will have. The occupation of the parents (farmers, teachers, nurses, industrial workers) and the nature of the family structure whether it is nuclear or extended contribute to this invisible organ that will be a part of the infant. Another important modulator of human development is geography. In the same manner someone born on top of the world have larger lungs than someone gasping his or her first puff of air at sea-level, the microbiome one develops tends to be different depending where one is born. Infants in Canada have a propensity to harbour more *Firmicutes* (energy harvester) than infants in India. The latter region promotes *Bacteriodetes*, microbes known for their ability to metabolize complex carbohydrates. Further studies will undoubtedly unravel the precise biological significance of these variations.

The initial microbial population residing in babies will also vary with the environment they are living in or where they are born. A birth in a home will promote a different set of microbes than a hospital delivery. Family members and pets in the house are a source of this microbial transmission while the sanitized ecosystem in a medical facility will provide the microbial landscape that the babies will tap from. Babies growing up in a surrounding with pets tend to be less allergic to these animals. As we start engaging with the new world around us, our invisible partners will evolve accordingly. The toys we play with, the place we reside, the social interactions we have, the food we eat and the medications we take will all have an influence on our emerging microbiome. Although some microbial species may accompany us from the womb to the grave, our microbial ecology is constantly evolving and adjusting to the new environment we are exposed to on an on-going basis as do all our visible organs.

2.4 Like the Lungs, the Microbiome Mirrors Our Environment

Our muscle, our heart and our lungs all adapt to the activities we are engaged in as does our invisible organ. For instance, the lungs of individuals born in elevated regions of the world are larger than those who first see the world at sea-level. The diminished levels of oxygen associated with high altitude program us to have an increased lung capacity and oxygen delivery system. Babies born in Tibet, Ethiopia, and the Andes where the elevation is higher than 2,000 metres will have a blood system designed for the lower oxygen content prevalent in these regions. The first gasp of air that is laden with oxygen will guide the eventual anatomical features of the pulmonary vessels in the same manner the microbes (in the air) native to rural Africa will play a role in the establishment of the invisible organs of infants born in this part of the world. Tibetan babies on average weigh more than Chinese babies and have more oxygen in their blood and nitric oxide, the artery-dilator. These features aimed at compensating for the level of oxygen at high altitude are absent in low-land dwellers (Box 2.2, Fig. 2.2).

Box 2.2 Geography Dictates the Development of Our Organs

The place where we are born has a critical influence in the development of our organs. The lung is an interesting example; their size depends on where we take the first gasp of air. At high altitude where the atmospheric pressure is lower, the lungs are larger than at lower altitudes. Furthermore, due to decreased oxygen levels per volume of air, the delivery of oxygen throughout the body has evolved to be more effective. The lungs take in on average 8,000–9,000 litres of air per day that diffuses through the blood system via the bronchi and the smaller bronchioles having tiny sacs termed the alveoli. An average human has approximately around 600 million of these air-filled structures decorated by a dense network of blood vessels ready to grab the oxygen and transport it to various part of the body. In high altitude dwellers like in La Paz, Bolivia or Lhasa Tibet, the larger lung capacity is punctuated with more haemoglobin, the oxygen carrier in the blood. To promote a rapid oxygen uptake, these inhabitants at 4,000 meters produce more nitric oxide (NO), a blood vessel dilator and 2,3-biphosphoglycerate, a compound that hasten the release of oxygen from haemoglobin. It is not surprising that many of the fastest long distance runners are from the elevated regions of Africa. The geography-triggered changes enable these inhabitants to adapt to the air in the environment; the same can be said of the pigmentation of the skin. Melanin, the molecule that gives the skin its color decreases as people live further from the equator, hence a paler complexion. These visible manifestations of the geography one is located in are also exhibited in the development of our invisible organ, the microbiome.

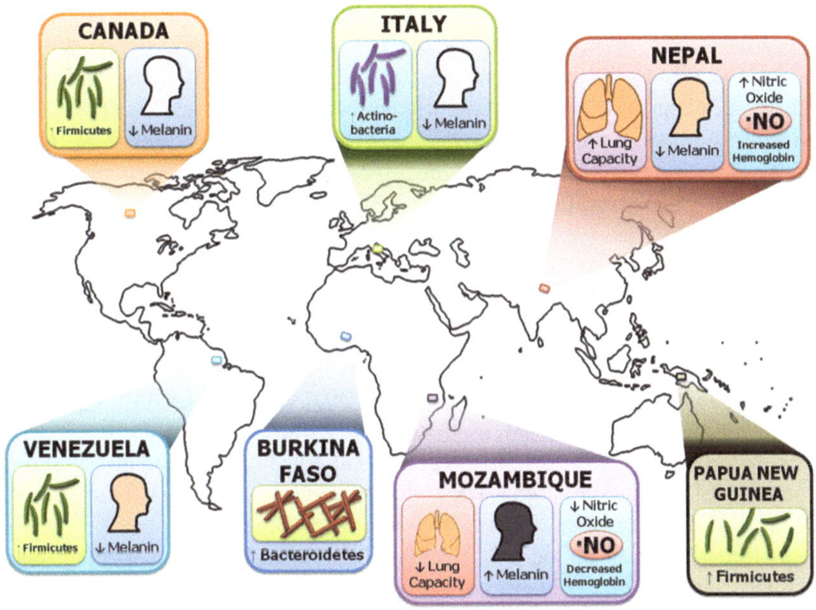

Fig. 2.2 Characteristics of numerous organs are a reflection of geography. (Lung capacity, skin color and microbiome constituents are influenced by place we are born and live in)

2.5 We Are What We Eat: The Microbiome – A Reflection of Our Nutrition

Food and medication are the biggest modifiers of the invisible and even the visible organs we possess. These factors have an immense influence on our microbial diaspora. For instance, the intake of antibiotics is a major disruptor of the microbiome in all age groups. However, babies and the elderly seem to be the most affected groups. Sustained and obsessive use of hygiene products can also retard and modify microbial colonization. These habits coupled with the food babies are fed will help shape a child's microbiome and give it the microbial ecosystem that will mould her or his initial years. Mother's milk is an important stimulator of select microbial communities.

2.6 Breast Milk – A Modulator of the Microbiome

Breast milk contains lactoferrin, an ingredient known to promote the proliferation *Bifidobacterium*, the live bacteria found in yogurt. This phenomenon is not observed in babies consuming formula or synthetic foods as these do not contain this *Bifidobacterium*-promoting ingredient. The formula-fed babies tend to have higher contents of microbes like *Prevotella* and *Lactobacillus*. Breast milk also contains the oligosaccharides that are aimed at recruiting *Bifidobacteria* in the colon of growing babies. These complex carbohydrates known as FOS (fucosylated oligosaccharides) are actually microbial fertilizers. They cannot be digested as humans do not have the necessary enzymes to do so. Hence, their existence in the mother's milk is most probably designed for the very purpose of acting as a prebiotic i.e. a food component that we are not able to metabolize on our own but that encourages the growth of good bacteria whose tasks are to help us accomplish various functions we are able to perform only with microbial assistance. They promote the colonization of the gut by a desired consortium of microbes. In fact recent studies have demonstrated that Malawian mothers with healthy children have an abundance of oligosaccharides (FOS) that are tagged with a unique product known as sialic acid. These FOS are readily broken down into simple sugars by the microbe *Bacteroides fragilis*, a prominent constituent of the invisible organ in healthy Malawian children. The microbiota-generated sugars subsequently become an excellent source of energy for the hosts. On the other hand, a group of post-partum Malawian mothers with limited sialylated milk tend to have undernourished children. It is evident that the special sugary component in the milk of these mothers is geared for promoting the growth of these unique microbes which then supply ready-made energy to the developing infants (Box 2.3). The magic of breast milk!!!. Indeed it contains a variety of other ingredients that are central to the quality of microbiome we will be saddled with for rest of our life (Fig. 2.3). It will not be surprising if soon regional variation in the breast milk components is discovered. These tailor-made products produced by the lactating mothers are there just to stimulate select microbes that provide a physiological edge to babies in a specific geographical location.

Box 2.3 Breast Milk: A Microbiome Nourishment

Breast milk contains a wide variety of nutrients like lactose, proteins and fats that are essential for the growth of the infant. It also has other bioactive components like vitamins, iodine, growth factors, hormones, antimicrobials and immune system activators responsible for the proper development of the baby. However, recent studies are pointing to the importance of milk in the establishment of the infant's microbiome. The complex carbohydrates in the breast milk are essentially foods designed to select microbes that help the infant during these formative months. These can only be utilized by the microbes but not the baby. For instance, these non-digestible sugars support the colonization of *Bifidobacteria*, the same microbes found in yogurt. They line the intestinal track and prevent the invasion by opportunistic microbes. They also fortify the immune system and in some infants high level of a species of this family namely the *Bifidobacterium infantis* have been associated with a larger thymus gland, an organ known to fight infection. The presence

(continued)

Box 2.3 Breast Milk: A Microbiome Nourishment (continued)
of hormones in breast milk is also an important contributing factor in the development of a healthy microbiome. Absence or limited intake of breast milk tends to result in the enrichment of *Clostridia*, a group of microbes known to be a harbinger of bad health. Breast milk is indeed microbiome food. This dictum should be heeded by all nutritionists in promoting good nutirition aimed at the microbiome. Eating for our microbiome should become a healthy mantra.

2.7 Solid Foods and Microbiome Enhancement

Once infants graduate to solid foods, they are confronted with another nutritional landscape that their nascent microbiome has to negotiate with and adapt to in the same manner the digestive tract does. Foods that require chewing lead to more of *Bacteroides* and *Clostridium* species. These microbes are responsible for the development of the gut. Here, the mucosal immune system of the intestine and the sentinels like peptidoglycan and lipopolysaccharide secreted by the resident

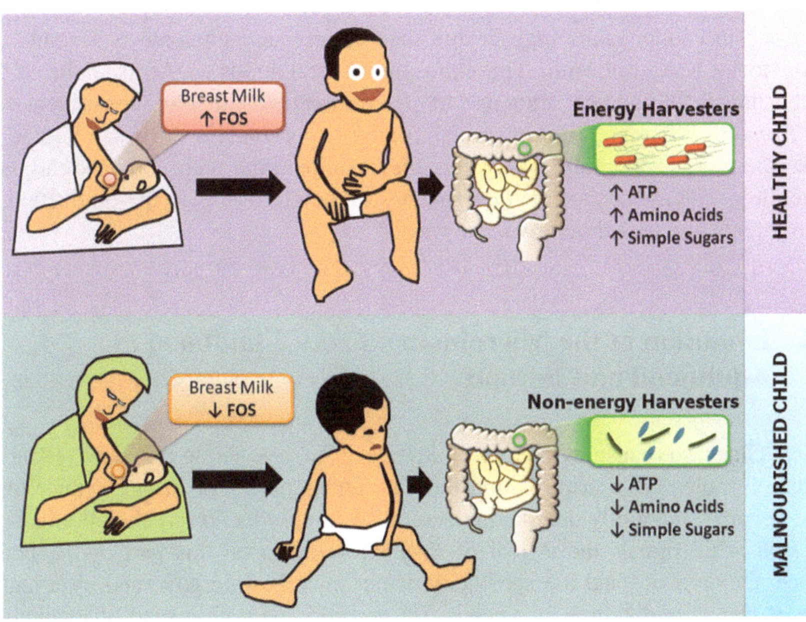

Fig. 2.3 The magic in breast milk. (Breast milk with fucosylated oligosaccharides (FOS) favours a unique microbiome dedicated to nourish babies; in the absence of FOS babies are malnourished)

microbes dictate the establishment of the microbial communities that are being established in this part of the body. The genesis of the microbiome is a process driven by multiple partners and necessitates input from the host, all the contributing invisible residents, the nature of food consumed and the biological changes the body is undergoing. For instance, the introduction of bananas and peas in the diet of a child tends to increase the *Bacteriodetes*, a family of microbes known to aid in digestion.

The nascent microbiome is reflective of very specific foods that are available in certain part of the world. Japan, a country surrounded by the sea is home to some unique sea-foods. Japanese consume sea-weeds as regular part of their diet. These products contain complex carbohydrates that are not readily digestible by humans. To mitigate this dilemma, Japanese children acquire microbes from their environment to allow them to make good use of these marine delicacies that would otherwise have stayed undigested. Someone from another part of the world does not have these specific invisible partners and hence cannot partake in these nutritious foods. That is why humans have different taste and affinity to foods depending where they are born. What are considered delicacies in a region may be just bland food in other parts of the world. Blame it on your microbiome.

We often experience this when we travel and indulge in local treats; this is becoming less of a problem as the global presence of the different food chains is providing us the same menus we enjoy at home. However, there may be a downside to this as we are unknowingly affecting our microbiome and at same time contributing to the world-wide decrease in food diversity we enjoy. A blending of taste due to eating habits may be not the best culinary outcome one wants the world to be bestowed with. The exposure to local foods in Africa induces the establishment of higher amounts of *Bacteriodetes* and less constituents of the *Firmicutes* family. This microbial configuration enables African children to make the most of the nutritional value they can get from their complex carbohydrate diets. The abundance of *Bacteriodetes* in the gut results in the production of energy fuels that would otherwise be wasted.

2.8 Evolution of the Microbiome: From Childhood to Adulthood and beyond

Others changes are also evident as children start to age. These are more reflective of the physiological transformation they are undergoing. For instance, their forehead becomes more oily due the increase in sebum production by the oil-glands in the skin. This signals the arrival of *Propionobacteria* on this part of the body. Hence, the diversity and the specificity of our microbes are governed by a multitude of factors and evolve as we age. The microbiome is in a state of maturation during the first few years of our life as are most part of the anatomical components (Fig. 2.4). For instance, the heart is only 25% of the size it will attain at maturity while the lungs will grow to full capacity nearly after 20 years we come on planet

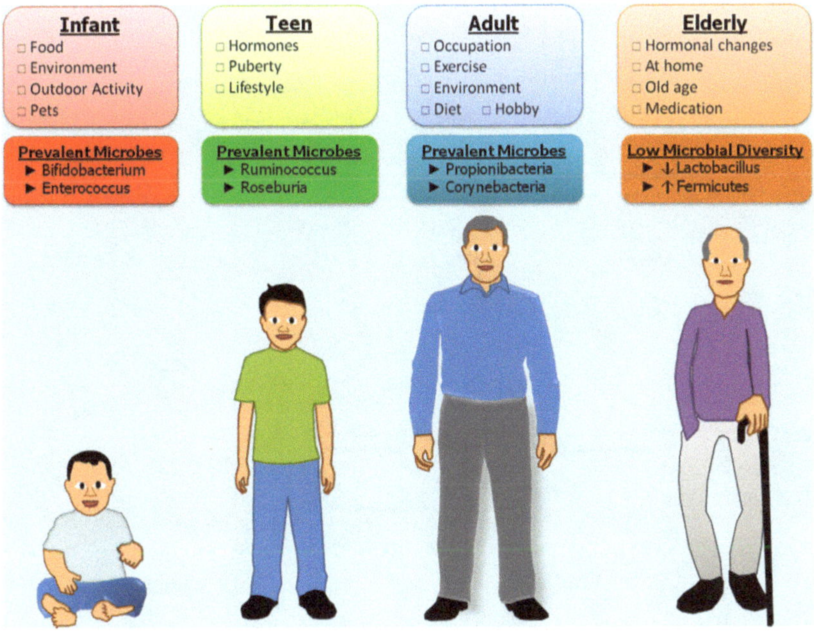

Fig. 2.4 The establishment and evolution of the microbome from childhood to old age. (hormones, life-style, medications and diet are some of the modulators; a decrease in microbial diversity is observed during old-age)

earth. Everyone is aware of the tribulations one has to experience when one's teeth start to protrude out of the gum. This process is very well-programmed and is very time-sensitive. The milk (primary) teeth commence to decorate our mouth after 14 weeks following our birth. The central incisor emerges unmistakably from the upper mandible, giving us that bunny rabbit look. It is joined successively by the lateral incisor and the canine a week or two later. These then undergo calcification that is then crowned and eventually rooted. The primary teeth are our oral guests for approximately 6–13 years. During this time the foundation of the permanent dentition is laid with the eruption of the central incisor to dot our oral landscape for rest of our life if we are lucky. The third molars or the wisdom teeth do not show up as late as 17–30 years. It is also important to note that this anatomical development process that we see happening on a day to day basis in our mouth is intricately regulated and is dependent on a variety of factors. Nutrients such as calcium, vitamin A, C, D and phosphorus are essential if we are to have effective dentition. Even what our mother consumes during pregnancy has a bearing on the formation of a child's tooth. The presence of chemicals like bisphenol A utilized in food packaging can have a deleterious influence on tooth formation. Good oral hygiene and fluoride-enriched water provide a boost to the maturation of our teeth. Here is an example of organ maturing that everyone has seen and can appreciate (Fig. 2.5).

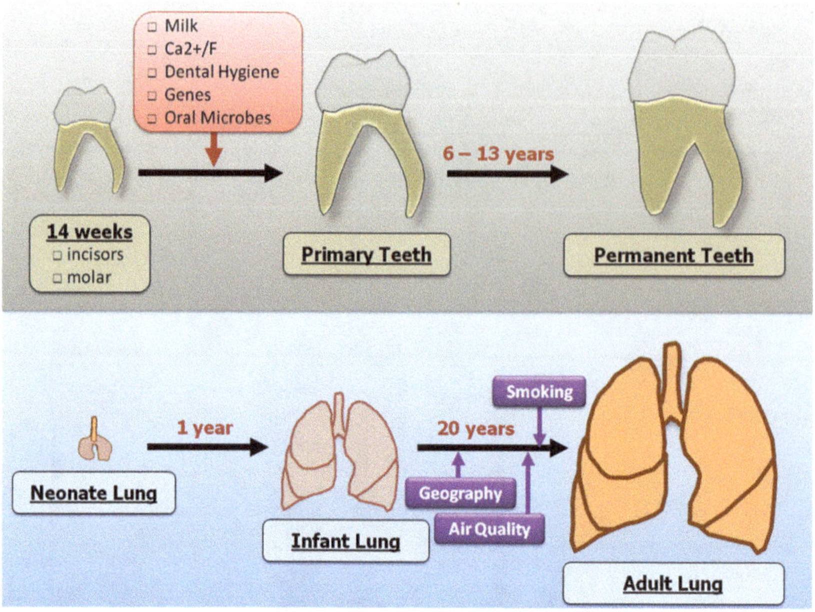

Fig. 2.5 The development of the teeth and lungs. (Just as the microbiome, the visible organs also undergo changes as age progresses)

This set the stage for a better comprehension of a similar phase that our microbiome has to endure. Unfortunately, we are visibly not privy to this as we are also unaware of changes, repairs and adjustments taking place in other traditional organs. Following the flux in the microbial partners our body experiences during the early years of our life, the microbiome tends to take a more permanent configuration after 3–5 years. The fingerprint of this microbial landscape can even be seen during our adult years barring major medical complications or change in lifestyle we may go through. Medical intervention has most drastic impact on our microbiome as does the constant use of medications. The exact nature of the microbial composition of each individual has yet to be unravelled and probably it will be sometime before the content of the invisible world residing within and on us is known. However, there is some guarded optimism that with the rapid advance of information and microbial fingerprinting technologies this will be possible sooner than later and the microbial profile of each individual will be a common routine diagnostic tool in patient care and more importantly in our well-being. Viruses, bacteria and fungi are all constituents of the microbial partners we possess. They collectively form our invisible body part in a manner analogous to our blood system with its white blood cells, red bloods and platelets etc. However, in this book only the significance of bacteria in this body part is being highlighted.

The microbiome of adult is basically made up of six main families (phyla) of bacteria. These are the *Firmicutes, Bacteriodetes*, *Proteobacteria*, *Fusobacteria*, *Verrucomicrobiota*, and *Actinobacteria*. The *Firmicutes* and the *Bacteriodetes*

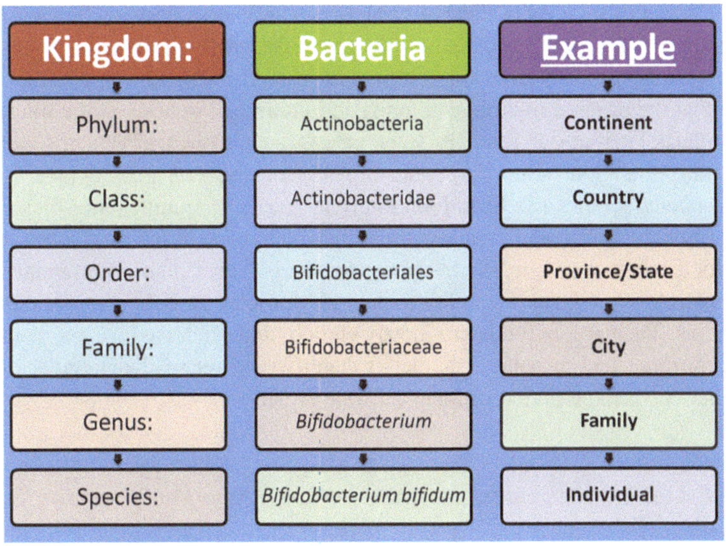

Fig. 2.6 Microbiome classification. (A comparison with the world and its inhabitants)

appear to be the most abundant ones. However, the subfamily and exact species tend to vary. If the microbiome is considered to be the world, the continents will be the families/phyla of microbes, the country will be the genus/genera, the identifiable group will be species and each individual will be the a specific microbe (strain) (Fig. 2.6). The complete mapping of an individual microbiome is currently a work in progress.

2.9 The Food Factor

The microbiome is a mirror image of the genes we inherit, what we eat, where we live and how we lead our lives. The same is true for all our body parts. Diet is an important modulator of the specificity and the relative abundance of various microbial species that roam the body although the initial foundation is laid by our genetic make-up and the vertical microflora transmission we inherit from our mothers (Box 2.4). While genera or species in the *Bacteroidetes* and *Actinobacteria* families can change, the population of these families is less susceptible to fluctuation during the life of an adult-it almost like the size of our major visible organs stay the same however the nature of its constituent cells change. As a general rule the amount of *Firmicutes* and *Proteobacteria* tend to go up and down during the course of our journey through adulthood. Individuals nourishing themselves on a typical western diet that is rich in protein and fats harbor higher quantities of *Bacteriodes*, a member of the Proteobacteria family. On the other hand, foods having an increased content of plant fibre promote an abundance of *Prevotella* in adults who are accustomed to including these in their daily diet. In rural Africa and some parts of South America, it is common to see high incidence of this microbial group in the adult microbiome. However, even in babies from these regions *Prevotella* is common. Hence, one can surmise that the original microbial colonization guides the genesis of the microbiome that manifests later in our lives. In some cases, unique micro-organisms have been detected in population raised on certain foods or in disparate geographical location. The *Treponema* species has been found only in rural communities in Africa as these microbes possess enzymes that allow their hosts to maximize the nutritional potential of the foods they consume. Some Malawian and Amerindian communities foster unique microbes that have the ability to degrade complex carbohydrates and generate amino acids. Their staple foods are mostly carbohydrate-rich cassava and corn. They usually consume less protein. This dietary habit would retard their regular development without the intervention of their invisible organ.

Box 2.4 The Microbiome is what we Eat: Nori-The Japanese Sushi Wrap

Nori is the Japanese term for edible sea-weeds. The most common of which belongs to the genus *Porphyra* that is widely used to wrap Sushi, a delicacy prepared primarily from rice and sea-foods. The species of the red-algae known as *Porphyra tenera and Porphyra yezoensis* harbour sulphur-rich carbohydrates that cannot be digested by most humans. However, the Japanese people who have been consuming these marine products for many centuries have been shown to possess a unique gut microbe with the ability to extract valuable nutrients from the polysaccharide, porphyran, an important component of the red algae. The bacterium *Bacteriodes plebeius* with its supply of the enzyme porphyranase can readily degrade this algal complex carbohydrate into galactose and sulphate, two ingredients that are readily utilized by the host. This microbe is not part of the microbial flora of humans whose diet is devoid of these sea-weeds. It is postulated that the gene for porphyranase has been transmitted to the gut microbe from marine bacteria living in sea-weeds.

2.10 Unique Food Breeds Unique Microbiome: The Hazda-Hunter-Gatherer Community

With assistance of their microbiome, they are able to barter the starch they intake with the amino acids that the unique microflora they have given refuge to supply them with. Starch exchanged for protein is a very lucrative arrangement for humans. If this was more common and widely distributed globally, human carbon footprint will indeed be lower, an attribute that would foster a less stressed environment. This will be sweet music to politicians who are grappling with rise of global temperature and climate change. The tripartite communication involving the host, the microbe and the environment illustrates how our invisible organ evolved in synchronicity with our body and the environment. This exact scenario repeats itself with our visible parts like the lungs and the blood content. Individuals who live on the top of the world where oxygen levels are lower than at sea-level have adapted to this atmosphere by having large lungs. Furthermore, their blood possesses a higher levels of 2, 3 biphosphoglycerate (2,3 BPG), a metabolite known to hasten the release of oxygen from hemoglobin, an attribute that allows the intake of more air. Well, the rationale of why we have this invisible organ is just in infancy and one can rest assure that more tantalizing discoveries will be unearthed and displayed in front our eyes. This knowledge will have major ramifications on our health, our planet and the way we perceive ourselves.

The Hadza, a hunter-gatherer community residing in Tanzania provides another neat example why we are on the brink of a major breakthrough on the significance of microbes within our body. Many had thought that all our organs we have, have already been uncovered ignoring the invisible organ that spans the whole body and constitutes the majority of our cells. It possesses unique microbial species that are absent in the people living in developed areas. The Hadzas' diets are composed mostly of plant-based fibres that are digested and metabolized by their very specific microbiome. They would have had much difficulty leading their current style if it was not for the unique invisible partners yet to be classified they have evolved with. The wealth of information on our microbiome pertaining to their diversity and their corresponding functions are waiting to be unravelled. It is important to note that globalization of the food industry may have a negative impact on the diversity of the microbiome, a feature we still need to explore and utilize for the betterment of society. Just imagine if we can intake foods that will be only the precursors of all the important nutrients we need. These will be generated and supplied to us by our invisible organ. Will this not be a game-changer of our consumer-based society to delight of so many humans who go hungry every day? And the positive impact on the global carbon footprint will be tremendous indeed.

2.11 The Maturation of the Microbiome: Influence of Chemical Exposure

Chemicals found in the foods we consume are also important contributors to the nature and specificity of the microbiome we possess. The intake of sulphite and sulphate-rich products favors the proliferation of sulphate-reducing bacteria. These ingredients are found in a variety of foods including alcoholic beverages, processed vegetables, deli-meats, and condiments. The increased population of the bacteria metabolizing sulphite and sulphate may disturb the fine-balance that governs the establishment of our microbiome and can result in abnormal biological functions. For instance, hydrogen sulphide, a product of sulphate metabolism is involved in the dilation of arteries. Diet laden with refined sugars and processed foods promotes *Clostridium difficile* and impedes the ability of *Lactobacillus* to colonize our intestine. The former is a major cause of concern and is known to be relatively resistant to antibiotics. Its infection in elderly is not easily treated. In fact, trehalose found in foods has been shown to promote the proliferation of C.difficile. In regions of the world where the use of processed foods is limited and complex carbohydrates are a common staple, *Prevotella* is predominant. In rural parts of developing countries this microbe can be readily detected compared to urban dwellers. This in turn is reflected in the health profile of these individuals. They are more prone to immune-related diseases compared to their rural counterparts. In general high microbial diversity in the microbiome is associated with better health outcomes than less diverse microbiome. Those who are healthy tend to have a higher gene counts while people saddled with unhealthy features have lower amount of genes associated with their microbiome. Recent studies have revealed that the adult microbiota can change rapidly depending on the nature of foods consumed (Fig. 2.7). A shift from a high fat/sugar diet to low fat/plant-rich polysaccharides can quickly result in a decrease in *Bacteriodes* and an increase in *Prevotella*. Hence, the diversity and the stability of the microbiome which are the crucial modulators of our well-being are responsive to the food we eat. The motto we are what we eat applies aptly to the microbial ecosystem we possess, hence the significant societal effort to eliminate the obesity triggering ingredients in foods. The edict 'eat for your microbiome' will become the operative rule once the significance of the invisible organ is taught more forcefully in health-related programs. Eating for our microbiome will undoubtedly lead to healthier outcomes for all. Chemicals in the environment also contribute to the microbiome profile we possess (Box 2.5).

Box 2.5 Pollution and the Microbiome

The pollution we are exposed to is a major mediator of the organs we are going to develop. For instance, the increased exposure to lead (Pb) severely affects neurogenesis and individuals living in areas polluted by this metal tend to have their cognitive ability diminished. Even the respiratory system is perturbed due to defective synthesis of heme, a vital component of haemoglobin the molecular vehicle that shuttles oxygen within the body. In a similar manner, the presence of pollutants in our midst perturbs the fine balance of the microbial communities that constitute the microbiome. Several species belonging to the *Firmicutes* family are affected by the presence of arsenic in the gut resulting in production of metabolites that impede the ability of the body to eliminate toxins. Cadmium is known to severely decrease the proliferation of *Bacteriodetes* that reside in the intestine and diminishes the production the short-chain fatty acid, butyrate. The latter is involved in a variety of functions including energy production and anti-inflammatory activities. Mercury, another metal toxin promotes the growth of *Actinobacteria* at the expense of other microbial community members like the *Bacteriodetes*, thus creating unhealthy conditions in the host. Exposure to organic pollutants like pesticides and polychlorinated benzene (PCB) do also modify the constituents of the microbiome.

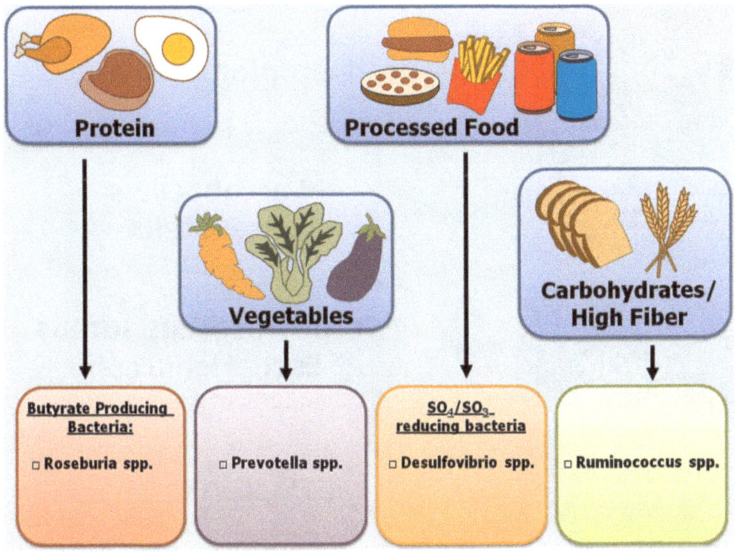

Fig. 2.7 Influence of foods on the microbiome

2.12 The Microbiome: A Reflection of Hormonal Fluctuations

Adults tend to have some characteristic finger-print that is reflected in their microbiome. However, the physiological changes they undergo in their lives have an important role in modulating their microbial ecosystems. Puberty, pregnancy, menopause and sexual differentiation in humans are marked by major biological transformations that are associated with unique manifestations in the microbiome. These very time-sensitive microbial partners aid us navigate through different stages of our life. During puberty the molecular responses triggered by the release of various hormones lead to anatomical and chemical changes that the microbiota has to adapt to. The emergence of hair on the skin is accompanied by increased activity of the sebaceous glands that help in oiling the hair follicles. However, this creates a greasy landscape on the skin that privileges some bacteria and deters others. For instance, more of the fat-loving *Propionibacteria* are found on the face, scalp and armpits. The release of male hormone testosterone and the female hormone estrogen also lay the foundation for the establishment of select microbial ecosystems in different organs including the gut.

2.13 Sex-Hormones and Microbiome

These microbial communities propelled by the sex-hormones are responsible of metabolic activities that favor fertility and reproduction. During pregnancy higher amounts of vaginal *Lactobacillus* and lower levels of *Prevotella* are reported

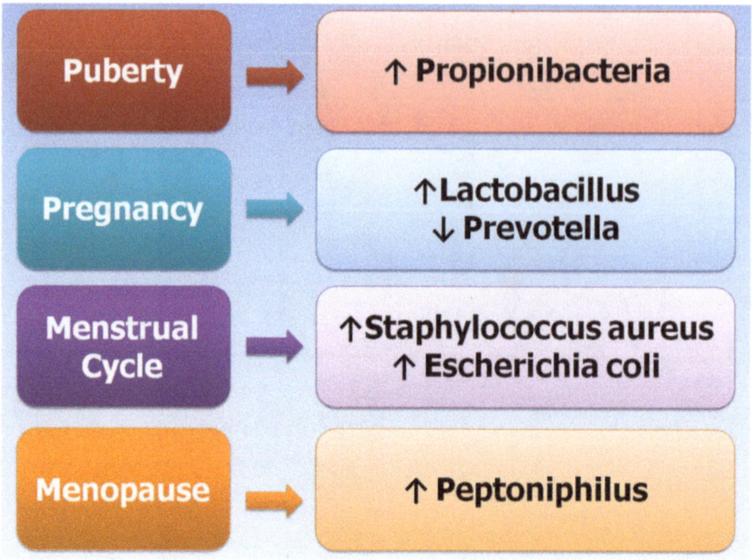

Fig. 2.8 Relationship between hormonal changes and the microbiome

compared to non-pregnant women. These microbes are important first-line defence system against pathogens. Armed with the ability to produce lactic acid, a chemical involved in maintaining a low pH, they dissuade the colonization of this organ by opportunistic bacteria. The relative stability of the microbiome during pregnancy is thought to be responsible for a successful outcome. The change in microbial dynamics in the vagina may have an adverse impact on this process. The microbial landscape is also affected during menstrual cycle. This occurrence is associated with an increase in *Staphylococcus aureus* and *E.coli*. During menopause, the decrease in the production of estrogen induces a major shift in the microbiome. The abundance of *Lactobacillus*, a hallmark of reproductive years in women is replaced by a variety of anaerobic microbes like *Peptoniphilus* that compete for supremacy in the vaginal community. This hormone-sensitive reconfiguration of the microbes is also seen in both the females and males. The level of testosterone affects the gut microbiome as this steroid signals the establishment of microbial communities that interact with the body defence mechanism. The low level of testosterone, a characteristic associated with femininity is reported to contribute to some autoimmune diseases due to its ability to affect the microbiome. Testosterone therapy is an effective remedy against these disorders (Fig. 2.8).

2.14 Life-Style and Climate: Impact on the Microbiome

The place we live, the kind of job we are do, the hobbies we engaged in and the climate we are exposed to all have a say on the microbes roaming in and on our body. Family members living in the same household share some similarities in the microbial ecosystems they nurture. If pets and plants are part of the community, exchange of microbiota is common. Practice of communal life-style as exemplified by certain tribes and cultural groups creates an environment for the swapping of our invisible partners. For instance, a Hutterite community has been shown to undergo a shift in microbiome during summer due to the availability of complex carbohydrates they consume together. There is an increase in *Bacteriodetes* and a decrease in *Actinobacteria.* In colder climates people tend to have more *Firmicutes* due to the presence of body fats while individuals living in warmer environments are host to more *Prevotella.* The microbial ecosystem of students living in the same residence with other students does exhibit some common attributes. The jobs we are engaged in to earn a living also add to our microbiome. Hospital workers, farmers and those employed in chemical plants usually have microbial population in their body reflective of those occupations. Farmers who are exposed to pesticides may contain chemicals that may dissuade the proliferation of some bacteria while microbiomes of those engaged in animal farms will undoubtedly strike a balance with the microbiota of pigs, cattle or horses they come in contact with. Similarly nurses are subjected to the microbial landscape existing in the wards they are working in and their microbiome will have to negotiate with this situation and will be reflective of this occupational environment they interact with on an on-going basis.

2.15 Animal Lovers and the Microbiome

Our microbiome adjusts and harmonizes with the surroundings we are used to. This may be the reason why some people feel sick while entering a hospital, a farm or a fish-canning plant. Our microbial communities are different from those individuals already exposed to such environments. One can become allergic to these work areas simply because our invisible organ does not possess the appropriate participants to help us cope with the new environment or our normal microbiome is disrupted by the chemical composition permeating these areas. However, inoculation with the microbial population that occupies these places help shift our microbiota in such a way that we get adapted if we keep encountering the same environment regularly. Hence, unknowingly, our microbial companions contribute to this process. Hobbies such horse-back riding, fishing, and hunting that abound in microbial activity impart lot to the microbiome of enthusiasts who practiced such leisurely activities. It will not be surprising if one day we find out that a horse-whisperer is only able literally to strike a conversation with a horse because he or she has been the proud recipient of a perfect dose of equine microbes during the numerous interactions with his or her four-legged friend. The regular act of feeding and patting the horse help in this exchange. Being in the same environment where there is constant intermingling of the microbial flora especially those residing in the lungs and the mouth provides ample opportunity for this mixing to take place in the two proponents. There may be some common microbial communities in both the horse and the rider, an outcome that further bonds them together.

2.16 Microbial Link: Pet Lovers and Exercise Enthusiasts

The same can be said of other animal lovers. In fact, adults who tend dogs or have dogs as pets have a higher diversity of microbes on their skin than adults who are not in contact with these canines. Regular exercise is an important modifier of our visible and invisible organs. In the same manner as going to the gymnasium can help tone our muscle, shape our body, improve our heart functions and allow our lungs to work effectively, this routine also remodels the constituents of our invisible organ. As we do not notice this microbial remodelling in the same we see our six packed-abdominal muscles appear, we tend to be oblivious to the microbial restructuring taking place. The change in the chemical gradients in form of oxygen, energy and a variety of metabolites that occur when we are running on the treadmill significantly help adjust and re-orient our microbial landscape. This exercise-conspired shift in our invisible partners is central to our health and well-being. Any spring allergy sufferer may be aware of the fact that their blocked nostrils are unblocked after a jog around the block. Any parents with kindergarden-attending children are acutely aware of the constant bombardment of infections during winter months due to the high density of microbes in an enclosed area.

For long we have taken this microbial support for granted simply because of our lack of understanding of how it operates and because it is not seen by the naked eyes. It keeps accompanying us discreetly. It adjusts aptly to the changing environment, to the food we eat, to the place we frequent, and to the physical activity we engage in. In fact, one can say that the microbiome is more in tune with us than many of our visible organs whose ability to respond may be limited due to the multi-layered command network they have to respond to. Almost everyone who travels to a distant land with distinct environment, and foods has experienced some discomfort in his or her biological processes. We may have allergies and hard-time partaking in the local delicacies. This is essentially due to a microbial landscape our body is not accustomed to. However, if our sojourn is long enough we tend to adapt as our microbial community incorporate some of the local invisible residents. These changes in the adult microbiome may leave an indelible imprint on our biological functions that may manifest long after we have left that particular geographical location. For instance, in the current global village we live in, it is not uncommon to be born in one country and to be living in another one with a complete disparate microbial landscape. A person born in the tropic and living in a nordic climate usually has the tools to fight tropical diseases that may not be prevalent in her or his adopted country. Hence, composition of the adult microbiome that is influenced by a variety of factors including climate, food, occupation, hobby, and our genetic make-up can also retain the remnants of the microbiota that we have acquired a long time ago (Fig. 2.9).

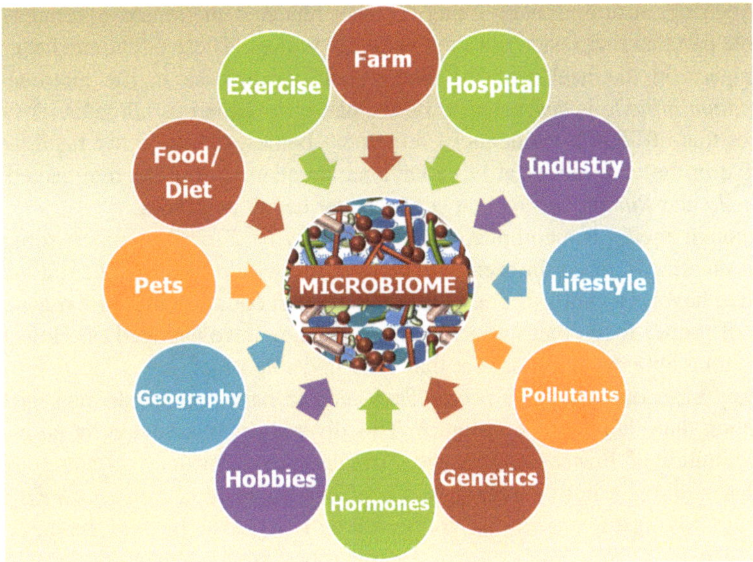

Fig. 2.9 Modulators of the microbiome in adults (occupation (farmers, nurses, industrial plants), pets, plants, hobbies, communal living, seasonal changes)

2.17 The Microbial Tango with Seniors and Centenarians

Adults aged 65 and above have lesser diversity in their microbiological flora. Seniors are a select group of individuals just like children who partake in unique activities that create an environment ideal for a major shift in the microbiome compared to adults. Although this does not apply to all seniors, a good proportion of them tend to live in retirement homes where they are in contact with a different set of microbial surroundings than that they are subjected to in a home-setting. The lack of variance in age of the residents they interact on a daily basis contributes to lesser diversity in the microbiome. However, in a house with various family members this is not an issue where all age groups intermingle. It is not surprising that some retirement homes are encouraging visits by school-aged children or teenagers in order to promote a normal life style-not being completely aware of the new invisible world these seniors are being enriched by.

Eden alternative home and the Green house project are novel concepts of senior living that are gaining traction in the USA. The inclusion of pets, gardening, children activities and regular visits by a diverse group of people is the hallmark of these organizations. Although psychological impact of these activities tends to confer a positive outcome in the well-being of the elderly residents, the microbial intermingling may have lots to do with this invigorating milieu. The limited mobility, a normal feature of retirement homes coupled with the regular intake of medications further compound the lack of variety seen in the microbiota. Poor nutrition and irregular consumption of foods are also reflected in limited diversity in the invisible partners that reside within and on the seniors. These situations have negative impact on the health of the elderly and an increase in the microbes like *Clostridium difficile* is suspected to be the cause of numerous ailments like bowel diseases that afflict the residents of retirement homes. On the other hand, seniors who live in the community at large have an abundance of protective bacteria like *Prevotella* and *Ruminococcus* that confer better health outcomes.

Although ageing is a complex process and may be difficult at times, some individuals are known to go through this phase of life with relative ease. And many of them live beyond a century. These centenarians can count on their good genes, the intake of proper foods and the active life style they have engaged in. However, it is becoming abundantly clear that that their invisible organ has a lot to say about why they have outlived their peers. They tend to have a more diverse microbial population than their elder colleagues. This diversity is punctuated by an increase in the families of *Proteobacteria* and *Firmicutes*. For instance, in some areas of rural Japan where a high percentage of centenarians resides, *Bifidobacteria* are an important driving force while in Northern Italy individuals beyond the age of 100 years have more *Ruminococcus*. These microbes are known to produce simple sugars that the host can utilize for energy production.

2.18 The Malleability of the Microbiome, A Highly Adaptable Organ

Even though the microbial partners accompany us from our developmental years to adulthood, to our senior years and beyond, the constituents of the microbial communities respond to numerous of factors that permeate our lives. Our mother's milk, the processed foods we eat, the soft drinks we drink, the sport we practice, the hobbies we are fond of, the places we visit, the social interactions we engaged in and the hormonal fluctuations we go through, all help in configuring the nature of the microbiome that becomes part and parcel of our body. Although these transformations are less evident than the changes we see in other body parts, they enable us to go through various ups and downs in life. We can see our muscle grow, we can experience a faster heartbeat when we run, we can see our greying hair but rarely we are even aware of the shift in the microbial population that governs various aspects of our life. In fact, the ability of the microbiome to rapidly respond to changes around us is essential for our well-being. Unlike, organs like the heart and liver that are made up of relatively same kinds of cells; myocytes in case of the heart and hepatocytes in the case of the liver, the microbial community comprises different cells including bacteria, viruses and fungi.

Furthermore this community has to comply with the rules imposed by the host and has to work in harmony with other residents, with their own genetic information and perform for the common good of all the partners. These demands imposed on all the community-members coupled with the distinct genetic commands they

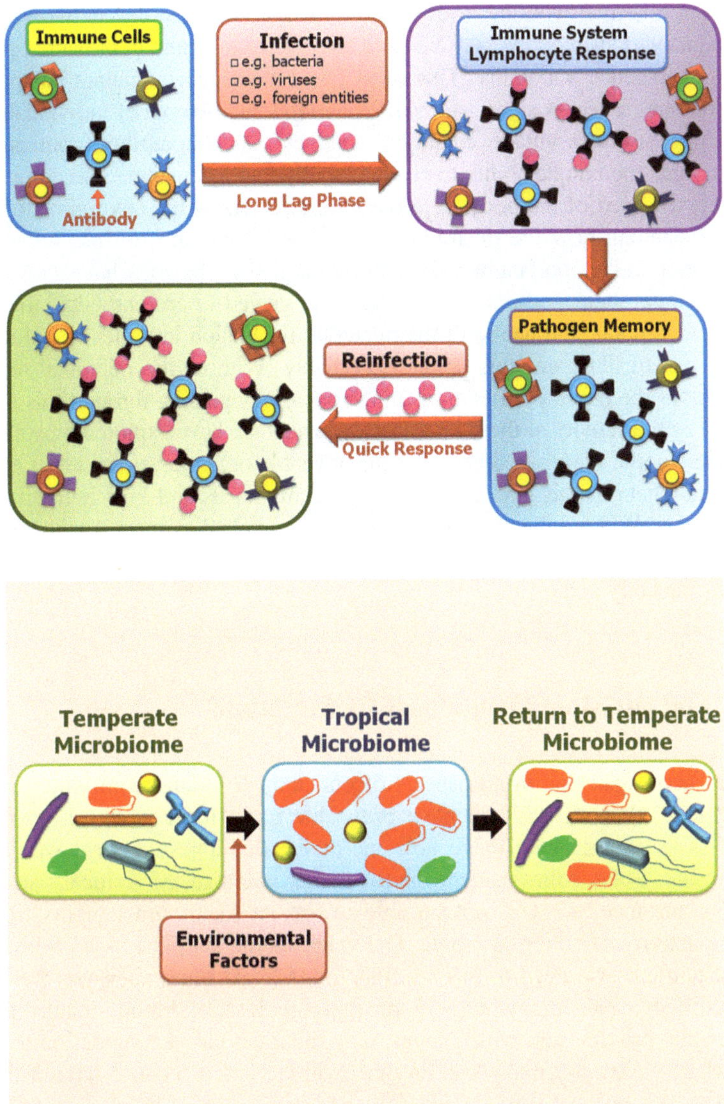

Fig. 2.10 Adaptability of the immune system and the microbiome. (As the immune system, the microbiome is also malleable; it retains some constituents of the environment it is exposed to; these systems have memory)

abide by ensure that the constituents of the organization are relatively malleable i.e. they can quickly evolved as opposed to the visible organs that are limited in this regard (Fig. 2.10).

The microbiome can be reconstituted depending on a variety of conditions while the same cannot be said of the heart. For instance, the basic anatomy and physiological functions of a normal working heart are more or less the same that is two ventricles, two atria and various blood vessels whether one is born in Nunuvut or Nantes possibly with some molecular variability. This quasi axiomatic statement cannot be applied to the microbiome. There is significant variation in the microbiome of individuals living in the same household, village, country or continent. Such an attribute confers on this invisible organ the unique ability to change rapidly and reconstitute into communities that reflect the need of the body in response to effectors in the environment. That is why understanding how the microbiome works will have an immense influence on our well-being. It is important to note that many organs in the body have evolved with the microbiome; their morphological features are geared to accommodate the microbiota and incorporate the nature of the microbial population into their functions. The digestive system illustrates this point very vividly. Without the microbial residents, this organ will be entirely different as numerous foods passing through this tract are processed by the activity of these microbes. Now that we have explored how the seeds of the microbiome are planted and how this invisible organ develops, let us examine the various functions this expansive organ performs for us and how it contributes to our well-being. Without the participation of these microbial partners during evolution, human beings would be an entire different creature if they had ever existed devoid of their invisible body part.

2.19 The Invisible Organ and Its Functions

Microbes accompany us in all aspects of our daily lives and contribute to a vast array of tasks that we will not have been able to perform if we are to rely on 'own' visible organs. Just image how inefficient our digestive tract would have been if it was not for the resident microbes that colonize this critical part of our body. For instance, we will not be able to extract maximum benefits from the mundane activity of eating an apple. Our microbiome supplies us with numerous essential nutrients we need to live on (Fig. 2.11). Next time when you have a cut and your body seals the wound to prevent the loss of blood, thanks the gut microbes that provide you with vitamin K, a must for blood coagulation to occur. Guess what! we cannot make this on our own, we either have to have it through the food we eat or have it made and delivered in situ for us by our invisible partners. Furthermore, the nutrients we will get from eating vegetables and fibres will have been limited and our pleasure in eating will be diminished if not for the microbes we harbour. In all likelihood, we would have to eat more to obtain the same amount of energy and consume huge amount of supplements like vitamins that we rely on our invisible friends for a generous portion daily.

But despair no more, our microbial partners that accompany us throughout our life journey see to it that we enjoy a variety of food products and obtain the most from our dietary intake. Indeed the digestive system is home to most of the microbes residing in and on our body. An adult contains approximately 2–3 kilograms of invisible lives scattered within and on the body.

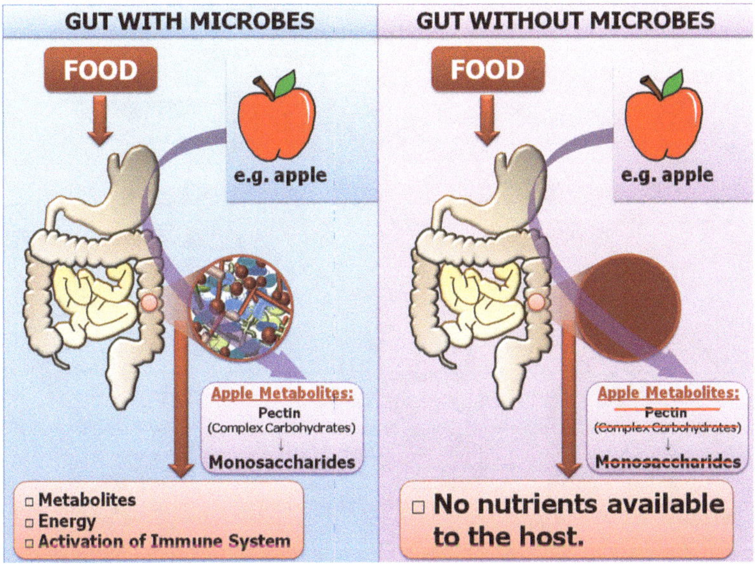

Fig. 2.11 Digestion of an apple with and without the microbiome

This is about 40 times bigger than the heart and 4 times than the liver, our largest internal organ. Since these important constituents of our body are dispersed all over and are invisible to the naked eyes, they have been taken for granted until recently. They are out of sight hence out of mind. Even though they are not as visible as the lungs or the eyes, yet they are pivotal to our well-being. Their contribution to our day-to day bodily functions outmatches the miniscule size of the participants. In this regard, our microbiome resembles the blood system with all its moving parts like the white and red blood cells crisscrossing most if not all of our internal nooks and crannies (Fig. 2.12). These blood cells guard us against foreign invasion and ensure we are flush with oxygen in all cellular components even those furthest from the lungs like the toes. The invisibility and dispersed nature of the microbiome should not preclude it from being a bona-fide and dependable organ. Indeed, it executes a variety of tasks that are indispensable to our well-being and we have evolved to be dependent on these microbes. The microbiome not only helps in digestion, it guards us against bad bacteria, heals our wounds, produces signals that communicate with our brain and helps our development (Fig. 2.13).

The microbes assist nearly all our major body parts in their daily activities. They roam the skin and reside in every crevice like the mouth, the nostrils, and the ears we have. The vast majority of the human microbiome calls the gastrointestinal tract their home where they instruct and aid our digestive system extract maximum nutrients from the foods we consume. Without their intervention, we will be not able to make vitamin B, a key participant in energy production, digest the bran wheat we eat and produce signals that warn us of any opportunistic invasion or quench our hunger – just to name a few (Fig. 2.14).

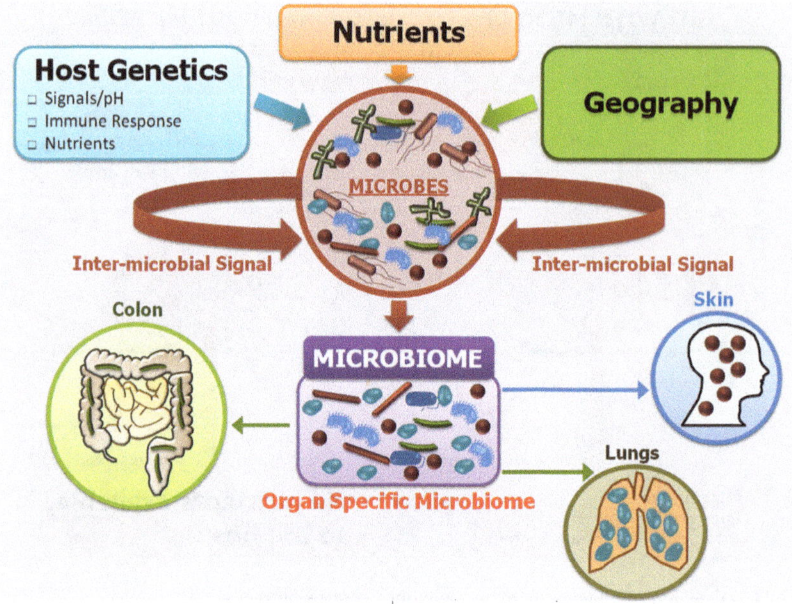

Fig. 2.12 The microbiome, an expansive organ like the blood system. (Different constituents synthesized by various organs constitute the blood system; the microbiome is made of numerous microbes that adjust to the commands and cues of all the participating members including the host)

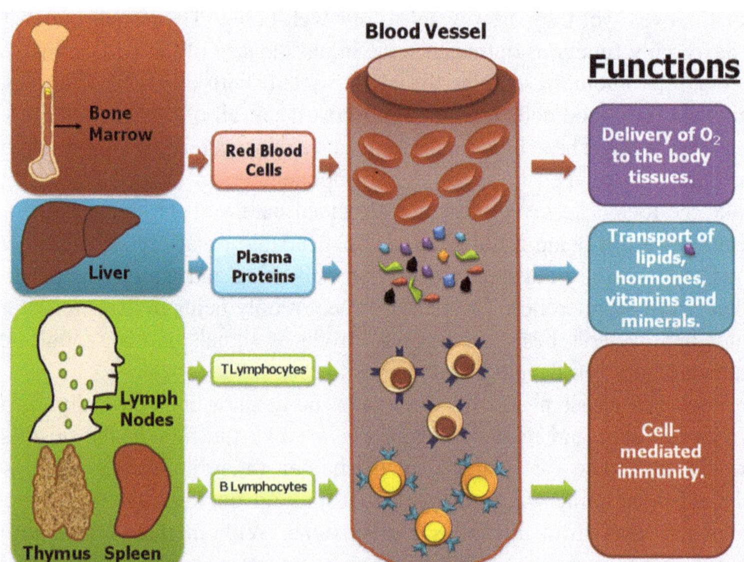

Fig. 2.13 The blood system with its components derived from the bone marrow, liver, lymph nodes, spleen, and thymus

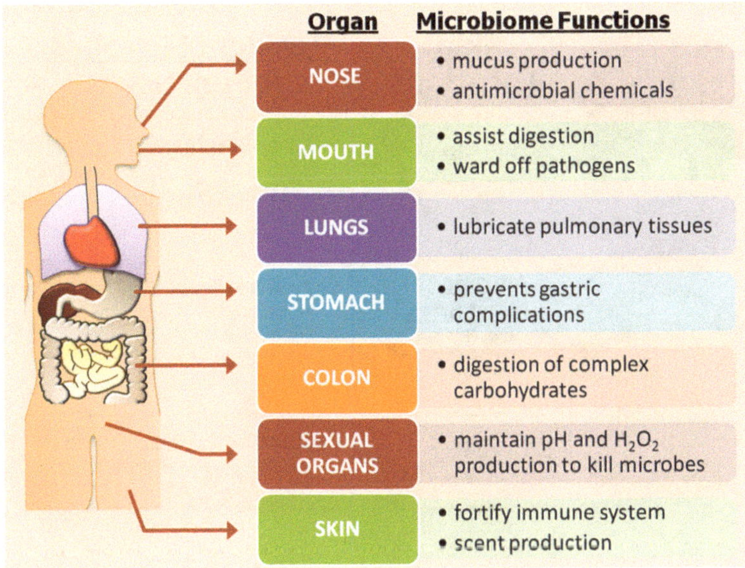

Fig. 2.14 Functions of the microbiome in different organs. (These organs have evolved with their resident microbes and will be ineffective without them)

2.20 The Skin-A Space to Roam

The skin is the largest organ of the human body and has a surface area of approximately 1.8 m^2 in an average adult. This extended area offers a diverse habitat that microbes have learnt to happily populate and adapt to. The architecture of the skin with its invaginations, folds, appendages, glands and follicles allow a range of organisms to flourish. This landscape is the primary physical barrier that invading foreign microbes encounter. The outer skin layer the epidermis, forms a formidable obstacle and resists intruders while maintaining the inside moist and full of nutrients. It is a virtual microbial zoo. As the first line of defence, the outer skin has evolved to utilize numerous good microbes to combat the bad ones. It has struck a partnership with some select microorganisms by providing them a home in exchange of their ability to prevent infectious organisms to invade the body. The unique topology of the skin and the availability of specific nutrients in select enclaves only favour the proliferation of the microbes that the body desires. This dermal landscape can be humid, dry, oily, salty, aerobic, anaerobic, warm and cool. Disparate combinations of these conditions can permeate this skinny abode. These microclimates permit only a distinct set of selective microbes to hug the skin. For instance, the groins and the toe web are partially occluded, tend to be warmer and humid while the elbow and leg are relatively dry and undergo temperature fluctuations. Hence, this diverse nature of the skin habitat promotes a wide array and density of commensal microbial life. The four main families of bacteria that dominate the human skin are the *Actinobacteria*, *Proteobacteria*, *Firmicutes* and *Bacteriodes*. The specific genera and species of microbe are dictated by the microclimates the skin possesses and the genetic composition of the host (Fig. 2.15).

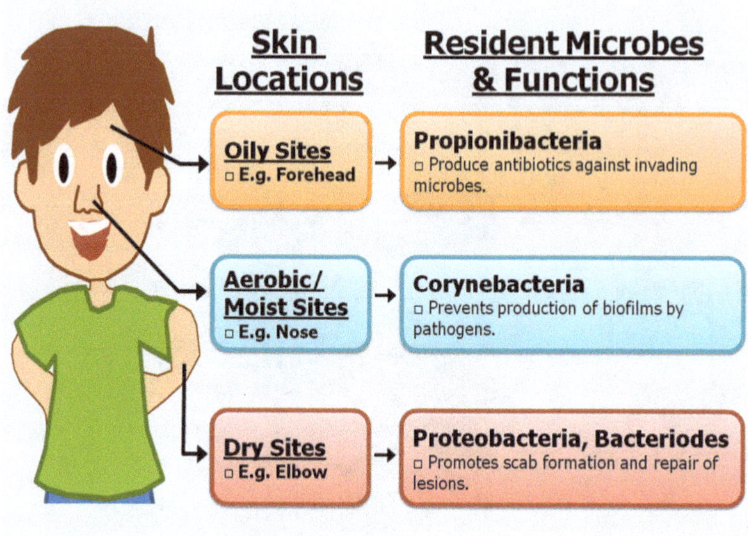

Fig. 2.15 The diverse microbial communities are adapted to the microclimates on the skin. They perform disparate duties

2.21 Distribution of Microbes on the Skin and the Tasks They Execute

The dry areas have a diversity of microbes like the *Proteobacteria* and *Bacteriodes* as opposed to the oily sites that provide refuge to essentially the *Propionibacteria*. The *Corynebacteria* and the *Staphylocci* residing in the aerobic and moist areas process the sweat into odour characteristic of each individual. In exchange of the nourishing environment the skin provides these microbes to live and thrive, they perform numerous chores that enable the human body to function the way it does. This microbial relationship is unique and has evolved overtime. It will be impossible for humans to survive if their skin was to be rendered microbe-free.

2.22 From Wound Healing to Fortification of the Immune System

The wonderful tasks these marvellous invisible magicians execute are numerous. They speed wound healing, prevent other harmful bacteria to reside on our skin and they fortify our immune system. The oil produced by the sebaceous gland under the hair follicle, is utilized by our invisible allies to generate a range of chemical arsenal like antibiotics against invading microbes and primers for the immune system. The microbe known as *Staphylococcus epidermis* prevents bad bacteria from attaching to the nostrils by blocking the production of their anchors known as biofilms. Without the biological ropes these bad characters are on a slippery slope as they sneak on to our tissues with the aid of these nano-cables. Their ability to set foot and establish a colony on the nares is thus thwarted. Wound healing is another key job performed by the dermal microbial ecosystem. This process is delayed in individuals suffering from diseases like diabetes. The high level of sugar (glucose) that characterizes the blood of diabetics favours the proliferation of opportunistic bacteria that retard the healing process. Yes, when you see your cut healing and a scab magically appearing, say thanks to your invisible organ, the microbiome. It is also responsible for the scent emanating from your body. With the assistance of the microbes, each individual has a characteristic odour profile. This is important for communication and survival especially in infants where the visual and mental recognition are not well-developed. A newborn depends on this scent fingerprint to identify his or her mum and other members of the family. That is why skin to skin contact between mothers and babies are recommended as this action help in scent recognition most probably through exchange of microbiota (Box 2.6, Fig. 2.16).

Box 2.6 Body Odor and the Skin Microbiota

The odour that emanates from the body is a product of the nutrients and the microbes on the skin. The sweat glands play a key role in maintaining body temperature by secreting sweat, a fluid containing numerous chemicals including water, NaCl, urea, bicarbonate, ammonium and lactate. However, the scent glands also known as the apocrine glands localized in specific parts of the body like the axillae (armpits) and the nipples of the breast release nutrient-laden fluid rich in electrolytes, steroids, proteins, vitamins and fats. These become an ideal feeding ground for the dermal microbial residents. *Corynebacterium* spp, *Propionibacterium* spp and *Staphyloccocus* spp are known to convert the nutrients into a variety of organic compounds responsible for the characteristic smell associated with each individual. The sulphanylalkanols like 3-sulfanylhexan 1-ol and small chain fatty acids like butanoic acid in varying combinations account for most of the odour in human armpits. The chemical composition responsible for the smell is dictated by the resident microbiota. An abundance of *Corynebacterium* spp is associated with smelly armpits while the presence of *Propionibacterium* spp elicits less malodorous scent. Thus far the worst odours have been traced to the microbe *Staphyloccocus horminis*. The transformation of steroids into androstenol and androstenone gives rise to musk-like and urinous odours. While propionic acid derivatives emanate a cheesy smell, the sweaty feet odour is due to isovaleric acid. Hence, the constituents of the skin microbiome are important mediators of the smell fingerprint each individual has to contend with.

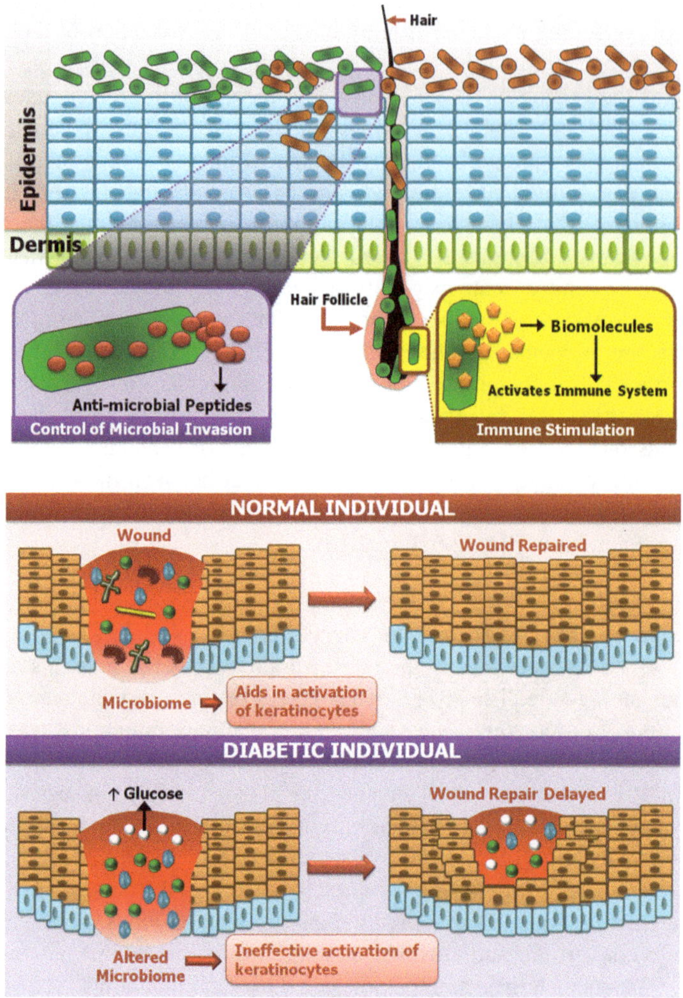

Fig. 2.16 Skin microbes are involved in wound healing; this process is compromised in diabetic patients due to high blood glucose content resulting in abnormal microbial community

The microbes inhabiting our skin play an important role in fortifying our defence system. Although the skin possesses an elaborate surveillance machine to guard us against harmful organisms in the environment, this task is supported and strengthened by our microbial partners. They produce signals like interleukin-1 that amply fine-tune the body immune response. In a way, they help the skin cells to select, calibrate and counter any undesirable encroachment by foreign entities. Thermal spas appear to play a role in promoting the growth of these immune-fortifying microbes. The curative power of some select thermal spas is attributed to this phenomenon. The natural chemicals found in these spas tend to enhance the microbial activity on our skin, thus allowing us to fend off diseases. Next time you look at your skin, think about the millions of microbes assiduously roaming on it as sentinels guarding you from the bad invisible characters in the environment. Your first line of defence is also made virtually unassailable only due to the fortification provided by our microbial allies. If you do not see them it does mean they are not there. They are working tireless for you. However, only if there is disturbance in this well-knitted community that their absence is felt in form of irritation or skin abnormalities.

2.23 The Invisible Soldiers in the Gut

Digestion is an essential aspect of our life as it enables us to extract valuable nutrients from the complex foods we eat. These nutrients are then utilized to fulfill an array of duties like generating energy, supplying the building block for growth and repairing damaged body components. The digestive tract where digestion occurs is a complex apparatus designed to maximize the nutritive value of the foods we eat. The nutritive elements are tapped in energy production and in providing all the ingredients for the body to run on. This task will be impossible without the intervention of microbes that shepherd the digestive process in such a way that the maximal benefits are squeezed out of dietary intake. In fact the digestive apparatus has evolved in association of the microbes it harbours. For instance, the colon of carnivores is simple, short and smooth as the undigested food is expelled relatively quickly. In herbivores, the same organ is long and complex as it is utilized for extraction of valuable components from the ingested foods, a process that will be devoid of any benefit without the presence of huge amounts of microbes. In humans, the long colon with small pouches that give it a segmented appearance is home to largest amount of microbes in our body. It is logical then to assume that the microbial residents had a say on how this part of our alimentary canal turned out. This partnership contributes to our ability to maximize the nutritional benefits from the foods we consume.

In an average adult, the complete gastro-intestinal system is 9 meters long and extends from the mouth to the anus covering an area between 250 and 400 m^2. This intriguing machine processes roughly 60 tons of food over the life-span of a human, a feat that will be ineffective and almost impossible without the laborious activity of the huge amounts of microbes that call this human territory their home. Indeed, the majority of the human microbiome lives here. These microbes participate in almost all the stages of digestion and ensure that the foods we consume daily release optimal amounts of nutritional elements. Besides, aiding in digestion, they generate vitamins that we are not capable of producing on our own (Box 2.7). Next time your blood clots whenever you have a cut thank the microbes within as they supply you with the crucial vitamin k that catalyzes the clotting process. For your friends who are slim despite being addicted to foods, blame it on the microbes they have. The vitamin B and the signals, the microbes secrete are central to the accelerated metabolic rate your friends are endowed with. Nearly 70% of the energy that is required to run the colon, is generated by the fatty acids produced by the bacteria that call the colon their home. They also release numerous chemical messengers that dictate a variety of our physiological functions (Fig. 2.17).

Box 2.7 The Intestinal Microbiome: A Supplier of Essential Ingredients, Vitamins and Bile Acids

The microbes living in the intestine provide the hosts with numerous vital micronutrients that they cannot synthesize on their own. They have to rely on the resident microbial population or have to acquire these from food sources. The folic acid family of vitamins is one of these microbe-derived chemical that allow us to function properly. The water-soluble vitamin B members like B1 (thiamine), B2 (riboflavin), B3 (niacin) and B7 (biotin) are generated in the colon by *Bifidobacterium* spp and *Lactobacillus* spp. These organisms possess the genes to synthesize p-aminobenzoic acid, a crucial building block for folic acid. These vitamins are involved in producing and in storing energy in the body. Vitamin K that plays a key role in blood coagulation due to its assistance in generating γ-carboxy glutamate is also generated by microbial activity in the large intestine. Almost 50% in the daily requirement of this vitamin is supplied by the gut microbiota. The primary bile acids (PBA) are synthesized by the liver and are important for digestion and absorption of lipids. These compounds can be transformed into secondary bile acids (SBA) in the colon. Here, the colonic microbiota with their deconjugating and dehydroxylating enzymes help convert these into chemical messengers involved in the regulation of glucose metabolism and energy expenditure, biochemical processes that are the basis of diseases like obesity and diabetes. Intestinal microbes like *Ruminococcus* spp and *Roseburia* spp participate in energy harvesting as they are an important source of pyruvate and lactate two metabolites that the host utilizes in the production of ATP, the universal energy molecule.

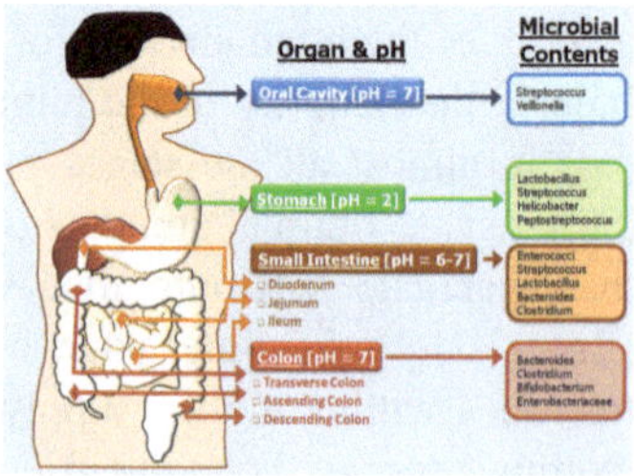

Fig. 2.17 Digestive tract with some specific microbes in different organs. (The stomach has the least amounts while the microbial population is the highest in the colon; Note the pH variations from ~2–7)

The five major families of bacteria that inhabit the digestive tract are: *Firmicutes*; *Bacteriodes*; *Actinobacteria*; *Proteobacteria*; and *Vermucomicrobia*. Thus far 50 genera and 1,000 species have been identified. This is a dynamic population and fluctuates with a variety of factors including diet, exercise and geographical location. For instance, when we travel we feel uneasy on arriving in a different country as our microbial system is in flux. It takes some time before we can adapt and acquire some of the local microbes which then integrate in our microbiome. It is more less how our body adapts to change in altitude, however, in this instance this process can take years compared to adjustment in a microbial community in the body.

2.24 The Mouth: A Major Gateway for Microbes

The first entry point to nourish all of our body parts is the mouth. It is composed of hard surfaces like the teeth and soft regions like the tongue and cheeks. This oral milieu provides a nutrient-rich environment for microbes to live and proliferate. The mouth has a dynamic microbial populace as it is almost an extension of the environment we live in. They secrete chemical soldiers that fend off invading bacteria and viruses that are in constant contact with the environment. Remember keeping your hand on your mouth while yawning diminishes the entry of rogue microbes – it is just not good manners, but this may protect you from invisible intruders that are always on the prowl for a safe refuge. These microbial guards release precise chemical sentinels that ward off parasites and contribute to the salivary immune system. They assist us during digestion and in eliminating foods that are not compatible to our well-being. In fact, the oral microbiome is on the frontline protecting us as it is the primary screening personnel of our body. The chewing of foods that helps increase the surface area for further digestive action to occur is primarily the domain of the teeth and the saliva. However, the secretions produced by the resident microbes contribute to this activity. Bacteria like *Streptococcus* and *Veilonella* are the harbinger of good health while microbes like *Granulicella* are abundant in individuals with dental cavities. Thus, the oral microbiota is an excellent guardian against opportunistic bacteria and the first frontier to ward off any unsavory foods. Remember the mouth microbial ecosystem also replenish the gut as some of microbes can make their way to this part of the body. The oral-gut axis is an important aspect of our digestive process. The chewed foods then proceed to the stomach.

2.25 The Acidic Stomach Is not a Microbial Desert

This organ has a unique acidic landscape that permits only certain kind of microbes to find refuge here. This inhospitable geography has few microbial dwellers but has become the home of the acidic-tolerant *Helicobacter pylori* and

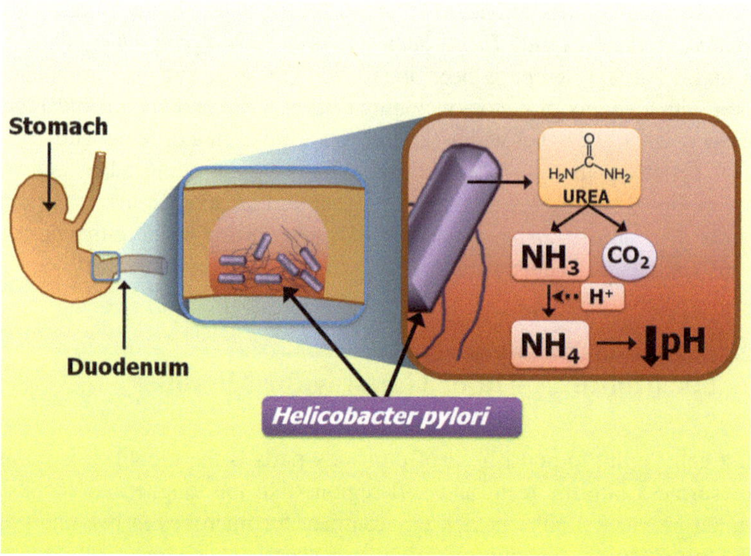

Fig. 2.18 Gastric microbiome survives in a harsh environment

some other bacteria like *Prevotella*. To survive in this harsh environment, these microbes stay close to the surface of the stomach, buried in the mucosal linings (Fig. 2.18). They have learnt to produce ammonia that aids in modifying the acidity of this organ, an act that prevents them to be nibbled away by gastric juice. Although their role is unclear, imbalance in the gastric microbiota may lead to diseases. Here the foods are acidified and mostly those containing proteins are chopped down into simple units before heading to the intestines where our microbial partners work on them with the aim of accomplishing tasks our own visible organ is unable to execute.

2.26 The Gut Microbiome

The adult human gut has the most bacteria and may be classified into six major families and numerous genera (sub-groups). Nearly 3,500 species have been identified so far. The *Firmicutes* and *Bacteriodes*, *Actinobacteria* and *Proteobacteria* family members outnumber the rest. Nearly a third of the gut microbes hail from the *Bacteroides* family and have a good amount of *Lactobacilli* (Firmicutes-family) and *Bifidobacteria* (Actinobacteria-family). However, the precise composition of the microbes in the gut varies with foods we consume, our geographical location, our occupation and most probably our genetic make-up. The large intestine tends to have a more stable and diverse microbial population than the small intestine. Dietary fibers constitute the primary nutrients of the microbes residing in the colon where they help break the food we cannot digest. The microbes extract

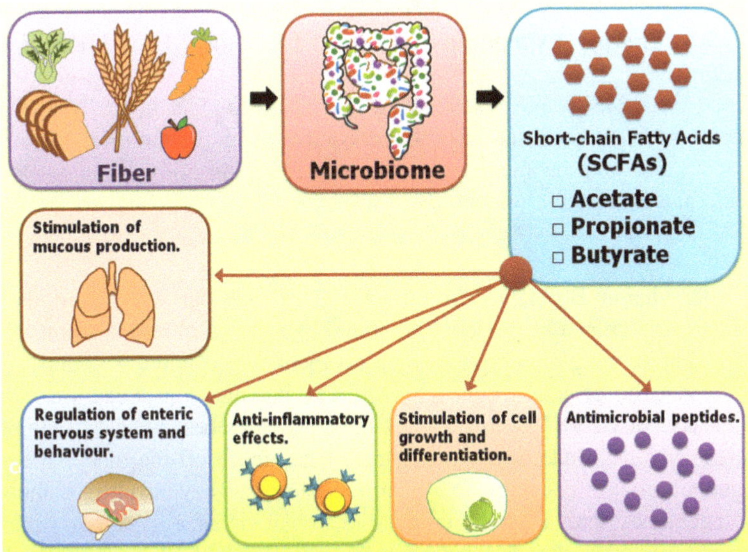

Fig. 2.19 Dietary fibre and short-chain fatty acids. They fulfill a plethora of duties and are critical to our health

energy from these nutrients and generate numerous metabolites that are essential for our daily life-style (Fig. 2.19). Our well-being is dependent on the laborious work these invisible partners perform. The simple nutrients like acetate, pyruvate and citrate liberated as a consequence of the microbial activity provide us with energy and control our appetite (Box 2.8).

Box 2.8 Energy Expenditure, Weight Loss and Gut Microbiome
Flavanoids are a group polyphenol compounds that are found in a variety of plant products including blue-berries, green-tea and grapes. Apigenin, a flavanoid abundant in parsley and maringenin, a flavonoid derived from grapefruit are known to play an important role in lipid metabolism and in the formation of adipose tissue, the fat storage in the body. Adequate supply of these flavonoids promotes the formation of brown adipose tissue. The latter is laden with uncoupling protein-1 (UCP) that mediates energy dissipation, a process resulting in weight loss. Although we obtain our flavonoid through our food intake, the intestinal microbiome regulates the amounts of these nutrients we are exposed to. Flavanoid-degrading enzyme like flavanone reductase that is expressed in numerous microbes like *Eubacterium ramulus* has the ability to decrease the amounts of apigenin and maringenin, thus triggering weight gain. Hence, intestinal microbes can be a cause of excess fats people accumulate. Secondary bile acids like lithocholic acid and taurolithocholic acid that are generated by intestinal microbes are involved in

(continued)

Box 2.8 Energy Expenditure, Weight Loss and Gut Microbiome (continued)

glucose metabolism and contribute to the energy budget in the body. The homeostasis of these metabolites that is controlled by the gut microbiota has major implications on the energy expenditure in the body, dictate our fat content and consequently a variety of diseases.

While the vitamin B obtained from microbes activates our metabolism, the unsaturated fatty acid known as linoleic acid contributes to anti-obesity and anti-diabetic properties. The priming of polyphenols makes them very conducive as anti-oxidants and the production of small chain fatty-acid like butyrate consolidate our armoury against cancer (Box 2.9). The activities of these gut microbes also dissuade opportunistic organisms to invade the digestive tract. They fence off the mucus membrane of the gut and prevent colonization by harmful bacteria. They also trigger the synthesis of antibiotics that augment the anti-microbial defence system of the host. Besides ensuring that the digested nutrients are absorbed, the microbes aid the potency of a variety of medications we take to combat diseases (Fig. 2.20).

Box 2.9 The Power of Butyrate, an Intestinal Microbial Product

Dietary complex carbohydrates are an excellent nutrient for the gut microbiota. As humans do not have the enzymes to digest these, they become important source nourishment for the colonic microbes that transform these resistant carbohydrates into short-chain fatty acids (SCFA) like acetate, propionate and butyrate. In fact, the concentration of these fatty acids may range from 20 to 140 mM depending on the microbial population, the fibre content and the intestinal transit time. They provide energy, prevent the invasion of pathogens, activate the immune system, and impede bacteria-induced preterm labour. Butyrate is known to modulate gene expression by inhibiting a key enzyme referred to as histone deacetylase (HDAC), a characteristic biological activity critical in the cellular development. It also has anti-inflammatory property due to its ability to inhibit the synthesis of cytokines like interleukin-12 and by eliminating sodium urate, an instigator of gout-associated inflammation. By helping augment the tight junction assembly due to its propensity to activate mucin synthesis, butyrate enhances the defensive features of the gut, an event critical for the establishment of a proper intestinal microbiome. While *Bacteriodetes* secrete acetate and propionate, *Firmicutes* are more avid producers of butyrate. Inulin-containing foods such as asparagus, leeks and bananas that are rich in fructose are some of the foods that microbes utilize to synthesize SCFA. The combination of microbial ecosystem and dietary fibres are central if humans are to get the maximum benefits from the intestinal SCFA.

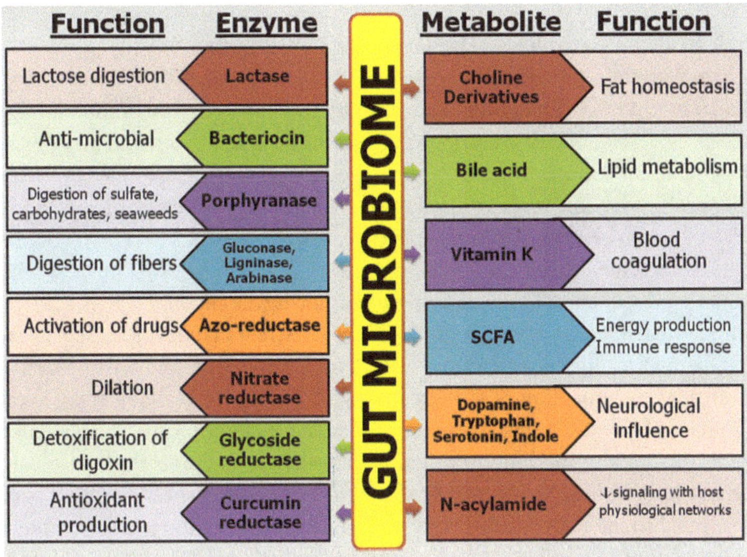

Fig. 2.20 Some metabolites and enzymes produced by the gut microbiome: physiological significance

For instance, the sulfa drugs that have both antibiotic and anti-inflammatory activity will not be effective if it is not for the gut microbes that help liberate their medicinal properties. Indeed, salicylazosulfapyridine, a medication aimed at inflammatory bowel disease (ulcerative colitis) is rendered potent by the microbes in the large intestine, precisely where it is needed. They are indeed our internal pharmacy (Box 2.10, Fig. 2.21). Thus it is quite evident that the microbes inhabiting our gut have co-evolved with us in a mutually beneficial manner. They accomplish numerous tasks that are critical for our well-being and our survival that our traditional organs cannot execute. Without these invisible accomplices, our digestive process and I dare say some of our physiological functions will not be the way we have come to know them. By divvying up some of the functions to this invisible organ, the digestive tract has been able to concentrate on some select duties. This arrangement has guided the components of the digestive apparatus to be anatomically the way they are. Hence, it is clear that our body parts will be quite different from the way they are in absence of the invisible organ we possess. Understanding this synergistic relationship is crucial to comprehending our own body and how it makes us tick (Fig. 2.22, Box 2.11).

Box 2.10 Gut Pharmacy – The Drugstore Within
Some of the microbial residents in the intestine are more or less like live pharmacists dishing out the appropriate active form of a drug or a natural supplement that the body utilizes to improve its well-being. They possess an array of enzymes that can modify biologically inactive or ineffective ingredients into

(continued)

Box 2.10 Gut Pharmacy – The Drugstore Within (continued)
medications ready to combat diseases or to fortify our health. These microbes can also bind to toxins and help eliminate them before they enter the blood circulation. Sulfasalazine is prescribed for ulcerative colitis. Its medicinal potency is dependent on the microbial enzyme azoreductase that releases the active form 5-aminosalicyclic acid with anti-inflammatory activity. Heterocyclic amines are generated when poultry, fish and meat are overcooked. These chemicals may trigger colorectal cancer. Lactic acid producing bacteria are known to bind noxious products before they can initiate tumor development. The effect of the cardiovascular medication digoxin is enhanced when taken in together in antibiotics aimed at eliminating *Eggerthela lenta*, a microbe known to inactivate this drug. The elimination of this species increases the bioavailability and hence the impact of the drug. The presence of oxalate in food products poses a problem as this moiety can bind to calcium and other free cations that can result in hyperoxaluria, kidney stones and renal failure. The microbe *Oxalobacter formigenes* found in the intestine has enzymes like oxalate decarboxylase involved in the degradation of oxalate. Natural products like curcumin (curry powder) and resveratrol (grapes, berries) exhibit anti-oxidant and anti-inflammatory properties due to the intervention of intestinal microbes that unleash these biological properties with the aid of enzymes like reductases and hydroxylases.

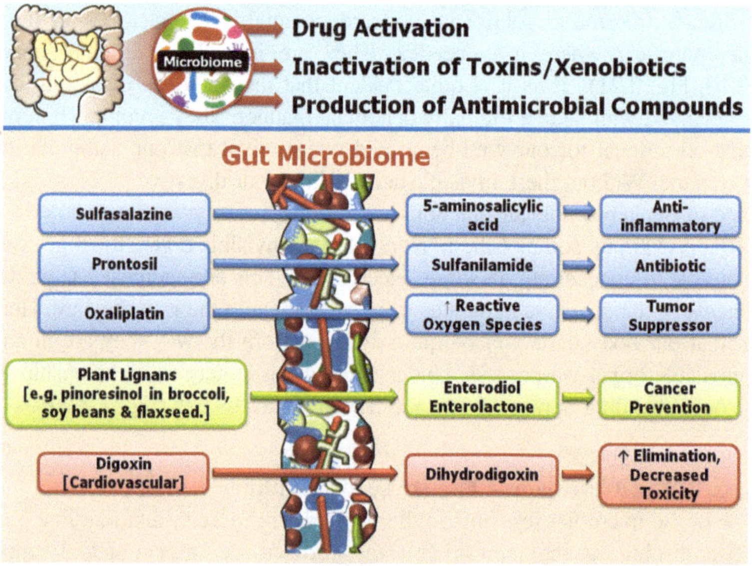

Fig. 2.21 The drugstore within the gut microbiome. (The microbes within the gut participates in a variety of medication-related activities)

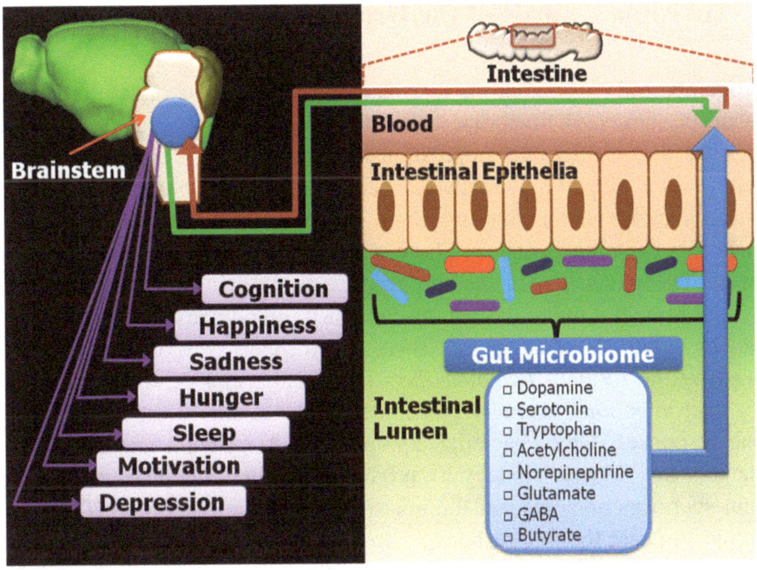

Fig. 2.22 Intestinal microbiome functions as mood modulators. (The neuroactive chemicals produced by these microbes can be communicated via the vagus nerve; the gradients of the neurotransmitters elicit various responses)

Box 2.11 Gastrointestinal Microbiota: A Factory of Neurotransmitters

The microbial population residing in the gastrointestinal tract is involved in the production of neurotransmitters that interact with the brain. These bioactive chemicals are derived from such essential amino acids like phenylalanine and tryptophan that are either obtained from dietary source or synthesized by the resident microbes. Tryptophan is the precursor of serotonin, a chemical messenger associated with sleep, appetite, memory and the release of gastrointestinal secretions. *Bifidobacterium* spp are known to produce tryptophan. Gut microbes can process phenylalanine into dopamine, a neurotransmitter responsible for behaviour, satisfaction, motivation and cognition. The level of gamma-aminobutyric acid (GABA), an inhibitory neurotransmitter with calming influence on the nervous system can be modulated by such microbe as *Bacteroides fragilis*. The vagus nerve that links the gut-brain axis is susceptible to the changing concentrations of chemical messengers originating from the microbial communities within, a situation that have major health implications. The ability of the microbial ecosystem to regulate the availability of these signalling molecules may be compelling us to work and behave according to the commands of our invisible partners.

2.27 Invisible Guards of the Urogenital Tract

The urogenital tract is another organ system that relies on its microscopic residents to execute a variety of functions. Unlike the digestive system, this organ varies significantly in the microbial population it harbors in men compared to women. Women have developed a mutually balanced relationship with the microbial world as microbes permanently colonize their sex organ. The *Lactobaccillus* species dominate this anatomical region. These microbes are mainly anaerobic, provide a natural barrier against pathogens and produce lactic acid, a chemical soldier that limits opportunistic microbial encroachment. This defense network is further fortified by other chemical guards such as hydrogen peroxide and bacteriocins produced by the resident *Lactobacillus*. These molecular sentinels shield against urinary tract infections, sexually transmitted diseases and assist the body's innate immune responses. Other bacterial communities *Atopobium and Prevotella* that complement this defense strategy by producing similar chemicals are designed to maintain such functions even if the microbial ecosystem is perturbed. They are a back-up just in case the first responders are disabled. The morphological characteristics of this organ make it a prime target of the opportunistic invisible intruders. The microbiome of the male genital tract is only now beginning to emerge and it is becoming evident that bacteria like *Lactobacillus, Streptococcus and Prevotella* are the important species that inhabit this territorial region of the body with biological attributes yet be unveiled (Fig. 2.23).

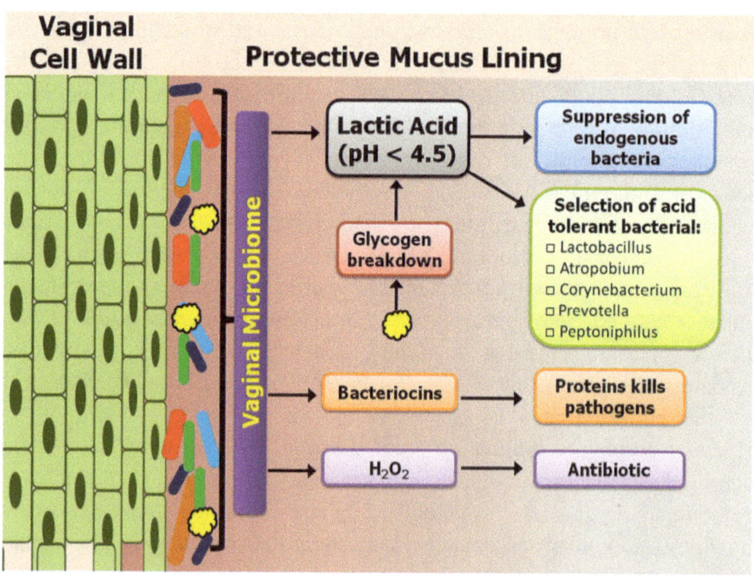

Fig. 2.23 Microbiome, the guardian of vaginal health

2.28 The Lungs: Home of Oxygen-Loving Microbes

The lung is confronted to a dynamic environment and is constantly exposed to the outside invisible world via the mouth and the nostrils. The movement between the lungs and the mouth is bidirectional as there is no clearly delineated physical barrier. The lungs are oxygen rich and tend to welcome aerobic microbes except in diseased situations where the air exchange is impeded. However, the temperature variation, the fluctuation in humidity and carbon dioxide along the pulmonary tract all contribute to the establishment of a diverse microbial population. The oral microbiome and the outside environment are the primary source of microbial entities that populate the human lungs. These microbial inhabitants tend to be dynamic and transient due to the intermingling of the external air and microbial content of the mouth. The pulmonary microbial ecosystem comprises mainly the families of *Bacteriodes*, *Firmicutes* and *Proteobacteria*. While the nasal cavity harbours *Moraxella* spp and *Streptococcus* spp, the lungs have a higher content of *Prevotella* spp and *Veillonella* spp. These microbial populations are suggested to play a role in easing the movement of air through the lung and most probably secrete mucus, a powerful lubricating agent that keeps this highly motile organ well-oiled. The microbes that establish colonies in the lungs are also like drill sergeants involved in instructing and fortifying the pulmonary immune system. It is being widely recognized that microbial ecosystem of lungs is critical for their proper functioning. Indeed the lungs of unhealthy individuals lack microbial diversity and are dominated by a single microorganism. This imbalance in microbial community is often observed in patients who suffer from pneumonia and cystic fibrosis.

2.29 Our Microbiome: An Indispensable Organ Like No Other

The microbiome is indeed an organ like no other organ we have. It permeates all aspects of our life as its constituents secrete a wide assortment of signaling molecules that have an impact in virtually all organs even those that do not even harbour any microbe like the brain. This invisible organ has been underappreciated and left to its own machinations for a long time. The advent of molecular microscopy and culture-independent techniques has enabled us to see this invisible organ in action in a relatively undisturbed fashion. These microbes can be witnessed performing their jobs and assisting us in our day to day activities. For the first time humans can admire their invisible partners in their natural ecosystem and welcome the sight that they are not alone but are being shepherded along the life journey by these tireless partners. From enabling us to extract the maximum nutrients from our foods, providing vitamins, fending pathogens and to rendering our body more suited to our natural landscape, our invisible organ performs a myriad of tasks that matches and even surpasses many of our more visible organs. The microbiome is highly adaptable and is the first responder whenever we venture in a new territory. It takes the lungs for instance years to adjust to high altitude if a

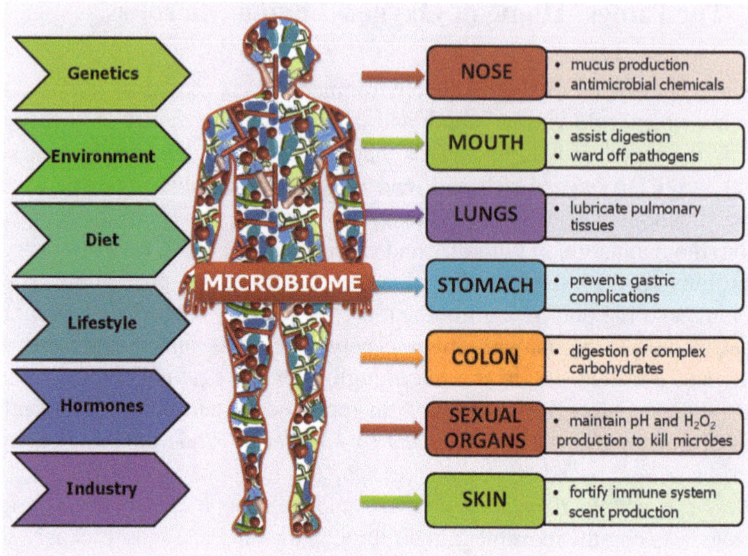

Fig. 2.24 Genesis of the microbiome and its functions

lowland dweller moves to higher grounds. But the microbiota remodels itself rather quickly with input from local microbial population. We have begun to appreciate this hidden body part as it is being laid bare with the assistance of new molecular visualizing technologies. The adage one can appreciate more what one can see really holds true in this instance. The dissipating veil of invisibility is rendering the significance of these microbes to our health and well-being impossible to ignore. No human being can exist without his or her microbial assistants. Our lives are inextricably linked. And from the very beginning of our existence these invisible helpers are with us guiding our development, maturation and our performance in this world. It is indeed a womb to tomb relationship, the intimacies of which have yet to be understood, appreciated and may be fine-tuned to the benefit of the host (Fig. 2.24).

Suggested Readings

Arnoldussen IAC, Wiesmann M, Pelgrim CE et al. (2017) Butyrate restores HFD-induced adaptations in brain function and metabolism in mid-adult obese mice. Int J Obes (Lond) 41:935–944. https://doi.org/10.1038/ijo.2017.52

Carmody RN, Turnbaugh PJ (2014) Host-microbial interactions in the metabolism of therapeutic and diet-derived xenobiotics. J Clin Invest 124:4173–4181. https://doi.org/10.1172/JCI72335

Charbonneau MR, Blanton LV, DiGiulio DB et al. (2016) A microbial perspective of human developmental biology. Nature 535:48–55. https://doi.org/10.1038/nature18845

Dinan TG, Borre YE, Cryan JF (2014) Genomics of schizophrenia: time to consider the gut microbiome? Mol Psychiatry 19:1252–1257. https://doi.org/10.1038/mp.2014.93

Dominguez-Bello MG, De Jesus-Laboy KM, Shen N et al. (2016) Partial restoration of the microbiota of cesarean-born infants via vaginal microbial transfer. Nat Med 22:250–253. https://doi.org/10.1038/nm.4039

Donaldson GP, Lee SM, Mazmanian SK (2016) Gut biogeography of the bacterial microbiota. Nat Rev Microbiol 14:20–32. https://doi.org/10.1038/nrmicro3552

Dorrestein PC, Gallo RL, Knight R (2016) Microbial skin inhabitants: friends forever. Cell 165:771–772. https://doi.org/10.1016/j.cell.2016.04.035

Fredrich E, Barzantny H, Brune I, Tauch A (2013) Daily battle against body odor: towards the activity of the axillary microbiota. Trends Microbiol 21:305–312. https://doi.org/10.1016/j.tim.2013.03.002

Graf D, Di Cagno R, Fak F et al. (2015) Contribution of diet to the composition of the human gut microbiota. Microb Ecol Health Dis 26:26164. https://doi.org/10.3402/mehd.v26.26164

LeBlanc JG, Milani C, de Giori GS et al. (2013) Bacteria as vitamin suppliers to their host: a gut microbiota perspective. Curr Opin Biotechnol 24:160–168. https://doi.org/10.1016/j.copbio.2012.08.005

Leung K, Thuret S (2015) Gut microbiota: a modulator of brain plasticity and cognitive function in ageing. Healthcare (Basel) 3, 898–916. https://doi.org/10.3390/healthcare3040898

Lin L, Zhang J (2017) Role of intestinal microbiota and metabolites on gut homeostasis and human diseases. BMC Immunol 18:2. https://doi.org/10.1186/s12865-016-0187-3

Man WH, de Steenhuijsen Piters WA, Bogaert D (2017) The microbiota of the respiratory tract: gatekeeper to respiratory health. Nat Rev Microbiol 15:259–270. https://doi.org/10.1038/nrmicro.2017.14

Morowitz MJ, Carlisle EM, Alverdy JC (2011) Contributions of intestinal bacteria to nutrition and metabolism in the critically ill. Surg Clin North Am 91:771–785, viii. https://doi.org/10.1016/j.suc.2011.05.001

Ravel J, Brotman RM (2016) Translating the vaginal microbiome: gaps and challenges. Genome Med 8:35. https://doi.org/10.1186/s13073-016-0291-2

Spanogiannopoulos P, Bess EN, Carmody RN, Turnbaugh PJ (2016) The microbial pharmacists within us: a metagenomic view of xenobiotic metabolism. Nat Rev Microbiol 14:273–287. https://doi.org/10.1038/nrmicro.2016.17

van de Wijgert JH, Jespers V (2016) Incorporating microbiota data into epidemiologic models: examples from vaginal microbiota research. Ann Epidemiol 26:360–365. https://doi.org/10.1016/j.annepidem.2016.03.004

Chapter 3
Dysbiosis, Probiotics, and Prebiotics: In Diseases and Health

Contents

Abstract The microbiome like any other components of the body undergoes numerous challenges during the life-span of a human being. These complications may involve injuries, aggression by pathogens, pollution, hormonal variations, genetic pre-disposition, unbalanced nutrition and onset of diseases. Although the microbial reconfiguration provoked by these stressors are not immediately evident as in the case of an afflicted visible organ where the abnormality is readily observable, the biological perturbations induced manifest themselves in form of various illnesses. The disruption of a working microbiome is referred to as dysbiosis and is a condition whereby the fine balance between the microbial communities and the host is distressed. Diseases such as cancer, irritable bowel syndrome, rheumatoid arthritis, acne, gastric ulcers, obesity and hypertension can ensue. The pathogeneses of some pulmonary disorders, digestive complications and neurological

V.D. Appanna, *Human Microbes - The Power Within*,
https://doi.org/10.1007/978-981-10-7684-8_3

abnormalities can be traced to the imbalance in the constituents of the microbiome. However, rebiosis, the re-establishment of the native microbiota is proving to be an excellent remedy against this condition. Probiotics, prebiotics, and synbiotics are potent therapeutic tools designed to rectify this situation. Probiotics such as *Lactobacillus* spp are more or less like stem cells utilized to replenish and rejuvenate the microbiome while prebiotics like fructose oligosaccharides (FOS) are microbiome fertilizers akin to mineral supplements or energy nutrients aimed at promoting the proliferation of select microbes in the invisible organ. Synbiotics is a combination of both probiotics and prebiotics in a proper dosage aimed at remedying dysbiosis. The molecular understanding of dysbiosis and rebiosis will offer a very effective non-invasive means in preventing and curing diseases with probiotics and prebiotics. This will have a dramatic impact on our well-being.

Keywords Microbial imbalance · antibiotics · cancer · obesity · anti-oxidant · drug activation

3.1 What Happens When the Invisible Organ Is Distressed? The Disruption of Our Microbial Communities and Common Diseases

As our microbiome accomplishes a variety of critical tasks for us, it is clear that we will not be the way we are without the microbes inhabiting in and on our body. Without this microscopic support, our anatomical constituents and our diet would have been completely different. The diets of carnivores and herbivores are also dependent on the microbial helpers these animals possess. The cows we see grazing would not be engaging in such an activity if it was not for the microbe-laden stomach they have. Indeed, humans also have evolved to be reliant on microbes to carry numerous functions they are not able to execute without the assistance of their traditional organs. Without the presence of these microbes, human anatomy and physiology would have been entirely different; we would not have been biologically the way we are. Thus, it is not surprising that if this harmonious relationship is perturbed, major complications may arise with disastrous consequences to the bodily operation. In fact, the delicate balance among the microbial ecosystems needs to be preserved if the body is to function normally. The metabolically compatible members of this community work in harmony and as they are assigned a specific job that enables the community to function properly. This synergy breeds inter-dependence and a specialization of functions. The community forges a collective work ethic amongst all the stake-holders. Constant interspecies communication ensures that their activity is controlled and excessive growth is under check. However, perturbation of this fine microbial-balancing act can result in debilitating impact on the host.

The increase in some bacterial species with the concomitant decrease of others triggers abnormal signals that can be the harbingers of diseases. This situation is similar to the operation of our social communities. For instance, if there is an excessive amount of doctors and limited number of nurses and other health professionals, the medical delivery system becomes sick. The disruption in the microbial community is bound to create a shift in the families, genera or species of the resident microbes. This can be fatal or make the business of living unpleasant. Numerous diseases are known to be triggered by the disturbance of microbial population within the body. The imbalance of the microbiome is referred to as dysbiosis (Fig. 3.1). Diet, physical stress, exercise, psychological distress, radiation, flying, humidity, geographical location, age and medications including antibiotics have been shown to exert demonstrable change in the microbial landscape in the body. For instance, ampicillin utilization leads to a sharp decrease in the gut microbiome, while the antibiotic, cefoxilin perturbs the fine balance by promoting the proliferation of *Clostridium difficile* at the expense of other microbes. The amount of some *Lactobacillus* species is sharply diminished during flight and may lead to anxiety associated with air travel. The increased levels of catecholamine produced during stress trigger the growth of pathogenic microbes like *E.coli* and impede the proliferation of good microbes like *Lactobacillus* and *Bifidobacteria.* Farm workers who are exposed to pesticides suffer a similar diminution in *Lactobacillus* and *Bifidobacteria*. Exposure to metal pollution like cadmium limits the amount of micro-organisms belonging to the family of *Bacteriodes* that in turn is reflected in reduction of metabolites like SCFA (short-chain fatty acids), contributors to numerous pivotal functions in the body as we have seen before. These changes lead to various abnormalities (Fig. 3.2).

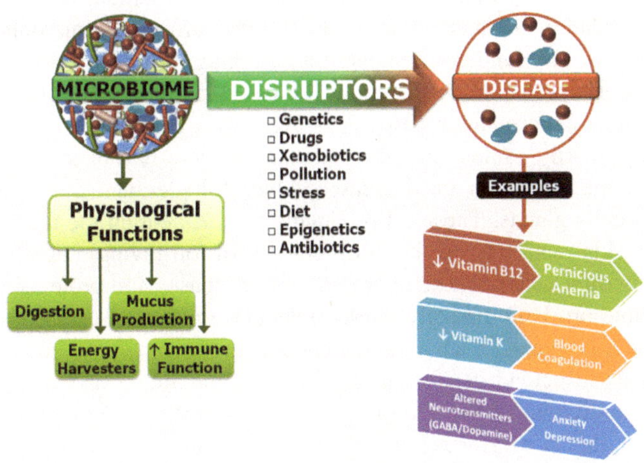

Fig. 3.1 Factors resulting in the disruption of the microbiome lead to dysbiosis. (Note the lack of microbial diversity in the distressed microbiome)

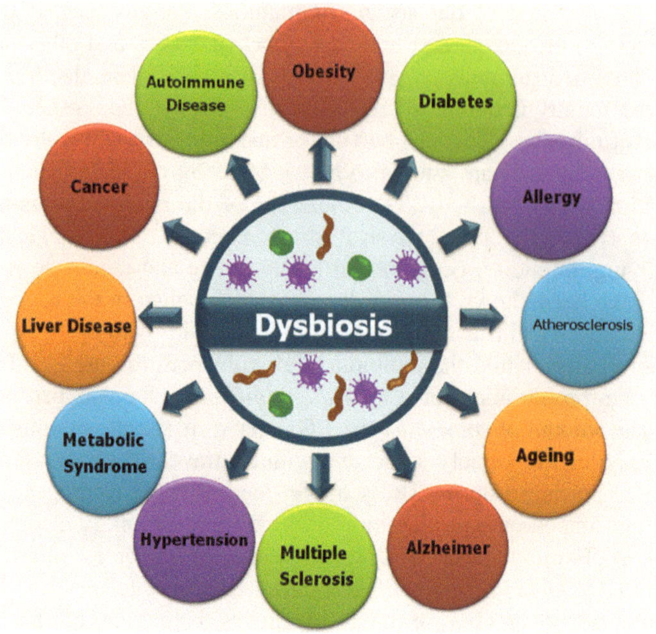

Fig. 3.2 Diseases and dysbiosis. (These medical complications can be initiated by dysbiosis)

3.2 Dermal Dysbiosis Leads to Acne and Eczema

The skin is the primary contact with the external environment and is constantly exposed to a variety of fluctuating conditions that affect the microbial communities residing on it. Humidity, temperature, clothing, cosmetics, soaps, age, and personal hygiene are some of disruptors of the skin microbiome leading to the onset of various abnormalities (Box. 3.1). Acne is a common skin complication that afflicts primarily adolescents. During puberty there is a major change in the landscape of the skin with the increase of hair follicles and the maturation of the pilosebaceous glands. These oil-producing vesicles enrich the nutritional content of the skin and trigger the proliferation of oil-loving microbe like the *Propionibacterium acnes*. These microbes secrete lipases and proteins that enable them to gobble oil. Unfortunately, these molecular scissors also bruise the tissues adjoining the oily glands. Such an unintended assault compels the body to unleash its own guards to defend itself, with devastating outcome. This counter response to the shift in microbial population results in the formation of those unwelcomed blemishes on the face. Hormonal imbalance may also aggravate this problem (Fig. 3.3).

Box 3.1 Disruptors of the Skin Microbiome

The dermal microbiota is prone to disruption by a wide variety of factors as it is permanently exposed to the external environment. Geography, pollution, ultraviolet radiation (UV-R), occupation and the host biology are major contributors to the skin microbiome. Injury, chronic conditions, infection, cosmetics and the use of hygiene products may result in a shift in the bacterial communities. UV-R promotes the release of anti-microbial peptides that may block the body's immune system and creates an environment for the cancer-causing viruses to flourish. Chemicals like triclosan, a common ingredient in household products like soap and toothpaste can perturb the fine microbial balance. This is further compounded by the inherent diverse landscape that the skin offers; the face is oily, hairy and exposed to environment while the axilla is occluded, moist and laden with microbial nutrients. The dysbiosis triggered by the host of environmental and microbial-driven features are at the origin of numerous skin diseases. For instance, the atopic dermatitis that is a chronic pruritic inflammatory disorder is characterized by an overwhelming increase in *Staphylococcus aureus*. A decrease in filaggrin, a microbial sensing protein coupled with a decrease in dermicidin, a peptide defending bacterial invasion results in this lack of microbial diversity and the proliferation of a select few microbes. Prebiotics like fucosylated oligosaccharides derived from chicory roots and probiotics like *Staphylococcus epidermis* are being introduced in cream to fight dermal dysbiosis.

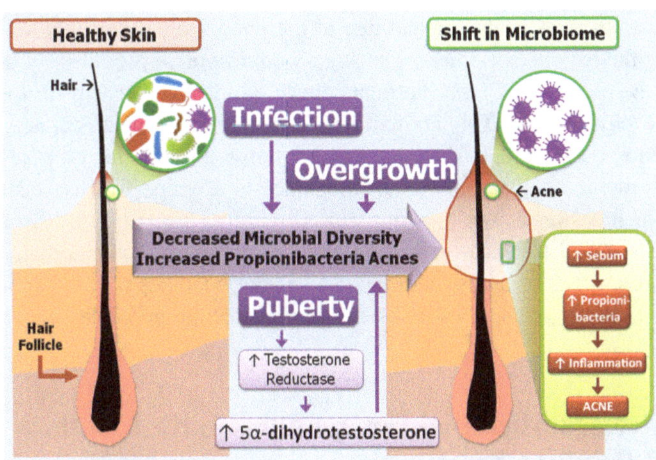

Fig. 3.3 Hormonal and nutrient fluctuations trigger acne. (The dysbiosis caused by these events provokes inflammation)

Eczema is another common skin ailment that finds its origin in the imbalance of the dermal microbial community. It is a chronic, recurring skin disorder that is more prevalent in children compared to adults. This disease is characterized by a major shift in the microbial population resulting in diminished biodiversity. Such a situation promotes the unchecked growth of the microbe, *Staphylococcus aureus*. Increased colonization by this organism provides a fertile landscape for the nasty *Streptococcus* to proliferate. The tight-knit functional community is unraveled. The new uncontrolled colonizers secrete toxins that induce the degranulation of the dermal mast cells and force the body's immune system to respond. This change in landscape is partly responsible for the dry red spots that are characteristic of eczema. These patches are sometimes accompanied by local inflammation.

In the case of psoriasis, a disease that is prevalent world-wide, similar conditions conspire to create a dramatic shift in the harmonious microbial population. The ensuing diminished biodiversity propelled by the over-representation of the genera *Staphylococcus* belonging to the *Firmicutes* family automatically leads to a reduction in other family members that had bonded into this dermal community and learnt to live in harmony. In this instance, the territorial presence of members belonging to the *Acetinobacteria* family is restrained. Thus, begins the initiation of psoriasis. Environmental factors like temperature, humidity or presence of cosmetics on our skin may hasten the onset and the severity of this malady. The establishment of the unbalanced microbial community residing in the psoriatic lesions generate antimicrobial peptides and other modulators that hyper-activate the body's immune system. This reaction modifies the life-cycle of the skin cells that respond by growing rapidly, a situation culminating in itchy, dry, red plaques. These become the hub of other uncontrolled microbial activity and can aggravate into psoriatic arthritis. Wound healing in diabetic patients is a problem as cuts on the skin of these individuals take unusually longer time to repair. Here again the inability to effectively recover from wounds is again due to the uncontrolled proliferation of certain microbes at the expense of others. The sugary environment provided by the diabetic blood and the presence of other nutrients create an ideal ecosystem for the multiplication of *Staphylococcus*. This nutrient-rich landscape on the skin-cut acts as a magnet for some bacteria and disrupts the harmony among the otherwise orderly community members. This microbial imbalance is a recipe for disaster and opens the body for invasion as some of our trusted partners are either too small in numbers to make any difference or are completely wipe-out of our intimate landscape by the opportunistic intruders. And wound healing becomes a very abnormal exercise for the body to partake in the absence of some of the invisible allies.

3.3 Distressed Microbial Communities in the Mouth: A Smelly and Painful Concern

The mouth is another part of the body that lends itself to constant contact with the invisible world. As it is an extension of our environment, the microbial traffic is high. Anytime we open our mouth to eat, drink, laugh, smile, yawn or cry, we are

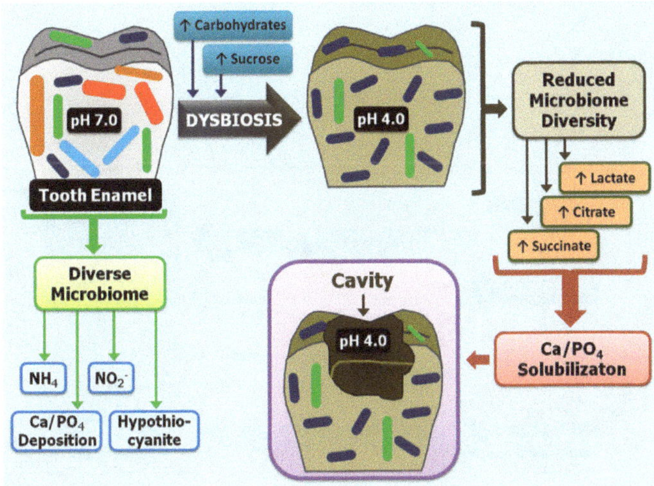

Fig. 3.4 Sugary diet and dental cavities. (The organic acids produced by the imbalanced microbial community decrease the pH and attack the enamel; hypothiocyanite is a potent antimicrobial agent)

enabling millions of microbes to access our body. However, due to the vigilant watch of resident oral microbes and their anti-microbial components, only a select few can contemplate to occupy a piece of the real estate either in our mouth or lungs. The microbial communities composed of diverse constituents that set foot are scrupulously and carefully interrogated by our molecular sentinels. Interactions with prospective residents are permitted only if they work for a common purpose. Only when a member can contribute to a specific function that the community depends on, the microbial member is accepted and a shelter is made available. The contribution of each of the member is critical for the functioning of the community.

However, this situation can rapidly change specially in the mouth where there is a constant flux of diverse exchanges occurring with respect to microorganisms. Furthermore the buccal chemical landscape also undergoes tremendous fluctuation due to the regular intake of foods and drinks. This situation can be complicated if one allows any food component to remain in the mouth for too long. Carbohydrate-rich nutrients become easy prey on which some opportunistic microbes can thrive. Despite the vulnerability of the mouth to possible invisible intruders, the microbial population is kept in check by the watchful guard of other community members including the host. For instance, in the establishment of dental caries, the *Streptococcus mutans* utilizes these sugary goodies stuck on the tooth to produce lactic acid, an event that increases the acidity of the buccal ecosystem. This presents an opportune situation for other microbes like the *Veillonella* and *Lactobacillus salivarus/acidophilus* to colonize this territory and further aggravate the affliction on the tooth (Fig. 3.4). The fine microbial balance that is perturbed as a result of ensuing chemical change in the mouth is an ongoing concern for people especially children with poor dental hygiene. This disruption and the resulting medical complication can be averted by controlling the chemical

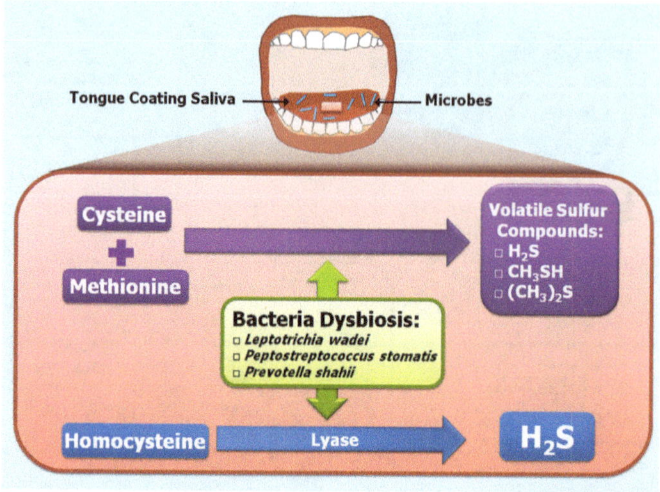

Fig. 3.5 Halitosis: bad breath and oral microbes. (The volatile sulphur compounds are the leading cause of bad breath, dysbiosis also impedes the production of antimicrobials like hypothiocyanite)

landscape in the mouth. Halitosis is also caused by microbial disruption that results in the increase in H_2S, a molecule responsible for smelly breath. This may also be aggravated by the genetic make-up of the host (Fig. 3.5).

The perturbation of controlled microbial growth in the oral cavity is known to trigger the gum disease referred to as periodontics. The homeostatic equilibrium in the microbial community for a healthy gum is maintained partly by the body's own immune system. This is upset by the invasion and colonization of *Porphyromonas gingivalis*, a microbe responsible for the destruction of gum tissues. These degraded components create a nutrient-rich environment that becomes a breeding ground for other opportunistic organisms that are usually kept at bay by our oral microbiome and the lack of appropriate nourishments they can thrive on. This feeding frenzy results in the overpopulation of some species compared to others and give rise to a misfit community of microbes. Bacteria such as *Prevotella* and *Desulfobulbus* are two culprits that exploit this situation and assist in the task of invading the gum. Thus, dysbiosis an uncontrolled microbial proliferation resulting from the temporary nutritional change in the oral ecosystem is at the root cause of numerous complications that ache both the hard and soft components belonging to this part of the body. The lack of microbial biodiversity in the crypts of the soft tissue like the tongue is at the origin of halitosis, an affliction that emanates unpleasant smell that most of us find repulsive (Box 3.2).

Box 3.2 Salivary Microbiome and its Health Impact

The mouth has the second most diverse microbiota in the body with over 700 species. The chemical landscape of the saliva plays an important role in the establishment of the microbial communities that reside in the buccal

(continued)

Box 3.2 Salivary Microbiome and its Health Impact (continued)
cavity. Nearly 10^8 microorganisms are found in one milliliter of saliva, a fluid rich in proteins, lipids, carbohydrates and enzymes. The presence of lactoferrin, lysozyme and lactoperoxidase act in concert to dissuade the colonization by fortuitous bacteria. Hydrogen peroxide, hypothiocyanite and nitric oxide derived from nitrate in foods are potent antimicrobial and are part of the defense armoury deployed to fend against cariogenic intruders. However, poor hygiene, nutritional habits, smoking and diseases like diabetes tend to weaken this defense. The presence of increased glucose level generates acidic metabolites that drop the pH of the mouth from 7.0 to 4.0. This promotes the acid tolerant microbe like *Staphylococcus mutans* and results in an increase in the *Firmicutes* family coupled with a decrease in the *Bacteriodetes* family. The solubilization of calcium and phosphate is preceded by a variety of dental diseases. Furthermore, the dysbiosis created by this situation leads to reduction of nitrate reducing bacteria, a situation restricting the output of nitric oxide, NO. This has major health impact including cardiovascular complications. The use of chlorhexidine containing mouthwash also has a negative influence on NO-producing bacteria. Hence, the saliva and its microbial residents are an important contributor to a healthy outcome.

3.4 Pulmonary Dysbiosis, Asthma, and Cystic Fibrosis

The mouth and the nostrils are constantly providing the lungs the air we need to live. Without this obligatory and mundane routine we will not be able to produce energy (ATP) that keeps our body machine active. This process is hard-wired in our brain that we keep on doing until we are no more alive. Thus we are constantly exposed to the microbes and chemicals in the environment where we are studying, sleeping, playing, working or just relaxing. The lungs are permanently bombarded with the myriad of micro-organisms our surroundings have to offer. Although the microbial populations in the lungs are well-maintained, these communities can be harassed by medications we ingest and by the chemicals in the air we are surrounded by. Farmer workers, employees in factories and other industrial organizations are at high risk in this regard as they are surrounded by chemical constituents that affect the microbial composition of the lungs. The development of asthma is triggered by the decrease in the diversity of the pulmonary microbial ecosystem.

The release of nutrient-rich mucus during lung infection aimed at arresting the proliferation of intruders can also inadvertently act as a magnet for some opportunistic microbes that further compound the problem. A shift in the microbial community favouring a group of *Lactobacillus* is observed during the onset of chronic obstructive pulmonary disease (COPD). During cystic fibrosis (CF) the viscous sputum allows the rapid growth of *Pseudomonas aeruginosa*, a microbe that is controlled in non-CF lungs. This microbial imbalance generates chemicals like alginate that further aggravates the situation. The nearly unimpeded pathway of the air tract between the mouth, lungs and the environment render these organs an easy target of dysbiosis. These body parts are challenged on an ongoing basis with changing microbial population and environmental factors such as chemicals and gaseous pollutants that make them a fertile ground for microbial disharmony. Hence, it is critical that the crevices in the mouth do not have nutrients stuck for long as these are perfect ingredients that can dislodge the well-established functional microbial communities and inflict us with diseases. The mouth and nose gears are a common sight in cities around the world with high level of pollution.

In Beijing, New-Delhi and Mexico City, citizens are often seen wearing these protective screens in an effort to limit the intake of the polluted air in order to diminish its influence on the lung microbiome. During outbreaks of microbial diseases such as severe acute respiratory syndrome (SARS) or influenza, this phenomenon is observed around the globe in order to ensure that the lungs are not further burdened by air loaded with viruses as these will undoubtedly promote dysbiosis. The mouth and lungs are the first line of defence to combat the growth of opportunistic microbes lurking around in the air. Thus, oral hygiene and the quality of air one breathes in will go a long way in helping maintain the proper microbial balance required for a healthy body. However, if these conditions are not met, microbial disruption resulting from the reduction of biodiversity in the pulmonary and oral landscape can be a cause of concern. It is not surprising that the deteriorating quality of the air across the planet has led to sharp rise in asthma and other lung-related illnesses (Fig. 3.6, Box 3.3).

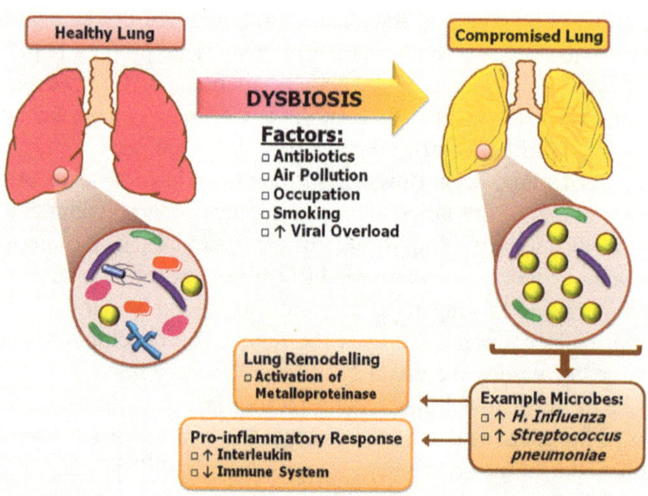

Fig. 3.6 Chronic obstructive pulmonary disease and the lung microbiome. (The remodelling of the lungs leads to decreased functional capacity)

Box 3.3 Microbial Link to Rheumatoid Arthritis and Asthma

Rheumatoid arthritis (RA) is a debilitating autoimmune disorder associated with the inflammation of the joints that can lead to bone erosion and deformity. The hyperactive immune response can also damage other parts of the body like the eyes, lungs and skin. This dysregulation of the immune system triggers pro-inflammatory tendencies with an increase in T-lymphocytes and self-reactive antibodies. Although genetic factors may contribute to this disorder, a major shift in the constituents of the microbiome is observed in RA patients. There is an overrepresentation in the *Provetella* species including *Provetella copri*, an increase in *Clostridium* spp and a reduction in *Bacteriodes* spp. This dysbiosis evokes the elevated production of homocysteine, hydrogen sulphide and lipopolyssacharide (LPS) that all known to promote inflammation, a feature culminating in the attack of one's own body part. Probiotics like *Lactobacillus casei* has been shown to mitigate symptoms associated with RA. Asthma is another inflammatory disease where the respiratory airways are challenged by the aggressive action of the immune system. There is an overproduction of mucus and an intense remodeling of the airway wall ensues. The presence of dust, pollen and spores tend to aggravate this situation. However, this disease is characterized by a marked change in the lung microbiota. In children there is reduction in such microbes as the *Faecalcibacteria*, *Veillonella* and *Rothia* while in adults an increase in *Proteobacteria* and a decrease in *Bacteriodetes* and *Firmicutes* are common. Hence, restoration of the microbiome with probiotics and prebiotics can be an alternative therapy.

3.5 Microbial Imbalance in the Digestive Tract: Gastric Complications

The digestive tract is the home of most bacteria in the body. Although the vast majority is housed in the intestine, some do find refuge in the stomach. Despite the inhospitable environment presented by this acidic organ, the microbes belonging to the *Helicobacter* spp have set-up their home here. They have done so by learning how to tame the low pH these environs are immersed in. They cling to the mucus of the cell-wall where the acidity tends to a bit more manageable. By neutralising this unfriendly territory, these microbes are able to develop a friendly habitat where they can reside. Even with these territorial adjustments, the stomach is the organ that harbours the least amounts of bacteria. However, the perturbation of this microclimate can quickly change this microbial landscape. The intake of medications like the proton pump inhibitors and antibiotics tend to promote a shift in this microbial population to the upper confines of the gastrointestinal tract. Once, this uncontrolled colony is established, it produces ammonia that compels an empty

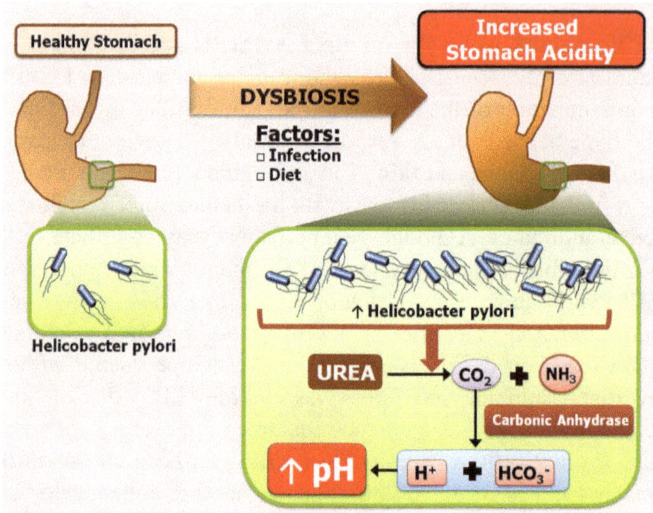

Fig. 3.7 Gastrointestinal microbiome and stomach diseases. (The disruption of the pH is the main culprit)

stomach to generate acid. It is critical that this digestive compartment is acidic only when it has food in its midst. Any change in the acidity in the absence of food distresses this organ. It is more or less like the stomach is chewing itself. This becomes the genesis of aches and heartburns that result into peptic ulcers. There is a rise in this microbially-induced disease around the world and the disruption of the fine working arrangement struck among the invisible residents may be at play. It is not surprising that the majority of ulcers can be cured by antibiotics and not by drugs aimed at diminishing the acidity of stomach, a finding that has dismayed many pharmaceutical companies as they have invested heavily on the incorrect cause of the disease. The understanding of how the disruption of these gastric microbial communities will pave the way for better remedies (Fig. 3.7).

3.6 Food Intake, Gut Microbiome and Diseases

The majority of the constituents of our microbiome is lodged in the small and large intestines with the latter harbouring the bulk of the invisible partners. Once the colonies are established in these locations, the stability of the residents is constantly being questioned due to the passage of foods and other edibles that transit through. This environment is under constant flux due to our daily intake of foods that can come in all shape, size, content, well-done, rare, or raw. The colonizers have to adapt. Low carbohydrate, high fat, high protein, dietary fibres, medications, antibiotics, sugary drinks all have major impact on the ecosystem. This constant pounding by these diverse ingredients, roughly around 60 tons during our

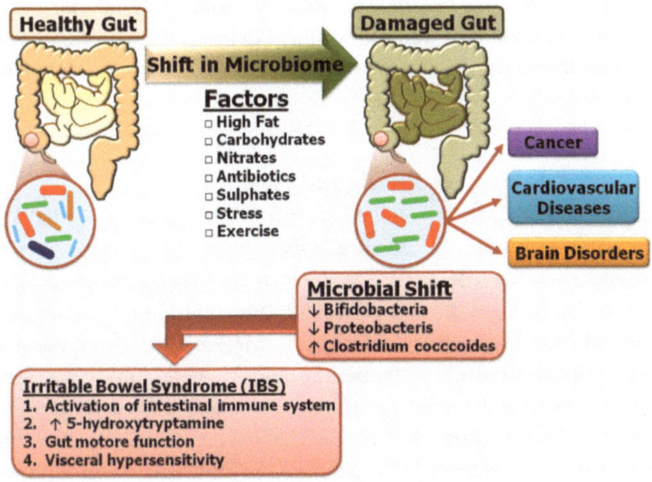

Fig. 3.8 Irritable bowel syndrome. (A reaction to microbial perturbation leading to inflammation and damage in the intestine)

life span is bound to take its toll on these miniscule residents within. Such a situation is akin to us changing our clothes conatantly in response to temperature changes in the environment. How would you feel if you have to face snow, rain, sun and fog on an ongoing basis and keep modifying your outerwear continually (Fig. 3.8).

Whether the food is starchy, oily or meaty; the microbes have to adjust. A shift from a high fat/high sugar diet to a regime of low-fat/plant-based polysaccharides can change the gastrointestinal microbial ecosystem in a matter of days. For instance, while *Bacteriodes* spp thrive in people who consume high fat foods, *Prevotella* spp dominates the guts of individuals who eat more carbohydrate-rich products; refined sugar intake on the other hand favours *Clostridium* spp. The lovers of vegetarian diets are host to less pathogenic bacteria while protein-rich nutrients tend to provoke increased activity of enzymes like nitroreductases derived from microbial sources. This situation contributes to the weakening of the inflammatory response and prevents the metabolism of short-chain fatty acids (SCFA). Remember the oxidation of SCFA like butyrate provides the intestines with an important source of energy and perturbation of this process can result in abnormal physiological functions. Preserved foods, dehydrated vegetables and alcoholic drinks with their high content in sulphate provide a fertile ecosystem for suphate-reducing bacteria like *Desulfovibrio* spp to thrive and distress the body. These micro-organisms generate metabolites that restrict the growth of probiotics that help us execute numerous tasks. Medications like opioids and the popular diabetic drug metformin impair the mobility of the gut. These are also responsible for shifting and deranging the community-like environment that good bacteria nurture. This disruption provides an opportune situation for bad microbe like the *Clostridium difficile* to multiply. Even psychological stress that is known to trigger the production of signaling chemicals like catecholamines has an impact on the microbiome. Well-established microbes like *Lactobacilli* and *Bifidobacteria* that are known to confer a number of healthy attributes in the body tend to diminish (Box 3.4).

Box 3.4 Food Additives, Dysbiosis and Health Impact
Numerous chemicals are introduced into foods to improve taste, to extend shelf-life, to add volume, to enhance appearance and to impart a variety of other properties. Some of these are known to interact with the microbial communities in the gut and perturb the fine communal balance. Dietary emulsifiers like carboxymethyl cellulose (CMC) is one of the additives that is widely utilized in foods, cosmetics, and hygiene products. Ice-cream, cookies, toothpaste and laxatives are some of the products where this polysaccharide provides texture, viscosity and structural integrity. Despite being mostly composed of glucose, CMC cannot be digested either by the visible or the invisible organs. However, it creates a major ecological change in the gut by interfering with mucin, a natural barrier lining the gut. Mucin is heavily glycosylated and act as a potent fence against any opportunistic pathogens to access the inner walls of the intestine. It inhibits the absorption of dietary flavonoids and also promotes the proliferation of microbial flora desired by the body. These functions are perturbed by CMC as it shares some chemical and physical features with mucin. This disruption results in dysbiosis leading to the multiplication of pathogens and their translocation into the intestinal tissues. There is an increase in *Proteobacteria* spp and inflammation ensues. Chronic intestinal complications have been attributed to these additives.

3.7 Microbiome and Global Epidemics: Obesity and Cardiovascular Diseases

Hence, it is not surprising that these chemical, environmental and behavioural changes that the gut microbiome has to contend with invariably lead to the disturbance of the delicate balance cementing the microbial communities and the host in an effective functional relationship. This flux in the gut ecosystem creates conditions for the abnormal proliferation of some members at the expense of others. The subsequent activation of enzymes and production of disease-causing factors can contribute to such maladies as irritable bowel syndrome (IBS), colorectal cancer, rheumatoid arthritis (RA), Type 2 diabetes, obesity, cardiovascular complications and neurological disorders. Some of the metabolic diseases like diabetes, obesity and heart ailments have reached an epidemic proportion globally. A recent report revealed there are more obese people than underweight individuals, a phenomenon that is a first for human history. An average person has become 1.5 kg heavier over each decade. Thus, if your grand-father was say 50 kg, your dad is 65 kg and you will be 80 kg at the same age of course. This scenario will be true if nothing else changes. The study on body mass index that was done over 40 years in 186 countries is alarming indeed. Policies on food and the quality of food we have access to will help to some extent. It is not surprising to see various North American cities are imposing an obesity tax aimed at sugar-laden soft drinks.

Examining the role of our microbial ecosystem is also pivotal if this issue is to be remedied. The gut microbiome is an important participant in the manner we digest food, extract the maximal energy and nourish the rest of the organs. A disruption in the microbial landscape can trigger metabolic diseases. For instance, during irritable bowel syndrome (IBS) microbes belonging to the *Firmicutes* family decrease while an increase in members of the *Bacteroidetes* family is observed. This disorder in the intricate microbial balance propels a concomitant rise in facultative bacteria like the *Enterobacteriae* spp. The unregulated establishment of the microbial community tends to impair the integrity of the mucus lining of the intestine. This shift perturbs the ability of the resident intestinal microbes to properly communicate with the mucosal barrier; such a situation results in the inability of the body to tolerate the harmless bacteria and neutralize the invading squatters. Following this distress imposed on this natural fence, conditions are perfect for the release of signals and chemical shuttles by the microbes to enable their migration across the intestinal defensive wall. This sequence of events forces the host to activate its guards with no clear mandate on the proper target and the result is an unintended inflammation giving rise to Crohn's disease and ulcerative colitis. These are the two most prevalent forms of IBS that can be traced to the perturbation of microbes residing in the gut. In fact, these diseases can be mitigated by increasing the presence of *Firmicutes* in the gastro-intestinal tract. It is clear that the conditions promoting the opportunistic microbes and the colonizing factors they produce are at the root cause of these disorders and re-establishing the original microbial communities goes a long in alleviating the pain of individuals suffering from these complications.

Obesity is a major problem afflicting the world. The rise in overweight individuals has been staggering and this malady attacks regardless of any economic and geographical boundaries. Almost all countries surveyed are experiencing the ill-effects of this metabolic disease. A variety of elements such as diet rich in simple carbohydrates, low in vegetables and fibres; life-style and genetic factors may be at play. However, the disturbance of the microbial harmony in the body cannot be discounted as a potential trigger responsible for the soaring rate of obesity around the globe. There is a clear distinction in the microbial landscape associated with obese people compared to underweight individuals. Lean individuals tend to have higher amounts of microbes belonging to the *Bacteriodetes* family while obese subjects and those suffering from metabolic syndrome are home to more members from the *Firmicutes* family. Although the specific species of these two microbial families responsible for the lean and obese traits have yet be catalogued, it is clear that obesity is characterized with increased levels of toxins like lipopolysaccharide (LPS) in the blood. Symptoms associated with weight gain and decreased sensitivity to insulin have been attributed to the presence of microbial toxins. Hence, modulating the nature of the gut microbiome and restoring the proper microbiota have to be included as part of the strategy dedicated to combating these disorders.

The food we eat is a major instigator of the establishment and fluctuation of the microbiome we possess. In fact, the adage we are what we eat applies aptly to the microbiota that constitute our body. One can even add that our physiological processes are the way they are because of the microbes we own. These invisible partners that accompany us throughout our live-journey are part of the elaborate communities responsible for our well-being. However, this is not a permanent relationship as it is susceptible to evolution and disruption by what we eat. This intimate relationship that we have forged with our invisible partners and all of our other bodily organs faces lots of twists and turns as we navigate through our lifelong journey. In the case of our microbial ecosystem these changes are non-stop as it is extremely prone to the vagaries of what we put in the mouth. And this is a daily activity that is most relentless and continual and commences the day we are born and ceased only upon our death.

Indeed it is important to remember that every component of body is in constant flux, we are never exactly with the same body parts that we are born with. These are evolving every second. For instance, we get a new skeletal system every 40 years and this happens on an ongoing basis day by day. Otherwise you will be soft like jelly one day and then wake up the next morning with a brand new skeleton. Thus, depending on our eating habits, we may be saddled with a microbial ecosystem that can promote or impede our wellness in the same way we have toned body if we exercise or we are stuck with a flabby belly if we are a potato couch.

A fat-rich diet together with a combo of a sugary drink is sure a concoction to upset the fine balance of our invisible organ. Such a nutritional habit is responsible for the non-alcoholic fatty acid disease (NAFLD) and cardiovascular complications. These greasy foods are known to promote the proliferation of some bacteria and arrest the growth of others. The establishment of this unbalanced ecosystem is a perfect brewing ground for the genesis of liver and heart diseases. These oil-laden nutrients contain copious amounts of choline, a chemical responsible for the yellow colour in the egg yolk. It plays an important role in the development of the brain in children. However, its excessive consumption via the food we intake specifically in adults can be problematic and the unwelcomed microbial residents aggravate our overindulgence in these choline-containing goodies. In reality our gut microbiota transform this into a product known as trimethylamine (TMA). This diversion of choline into TMA prevents the body from making some good lipids and subsequently leads to the accumulation of the bad lipid known as triglycerides. Remember the number 200 associated with triglycerides in the blood of adults. This magic number if higher is a relatively precise predictor of potential coronary abnormalities. These triglycerides have nowhere to go but to accumulate in the liver where they become the precursor of fatty acid disease. This is not all that can come out of this greasy food, it can become worst.

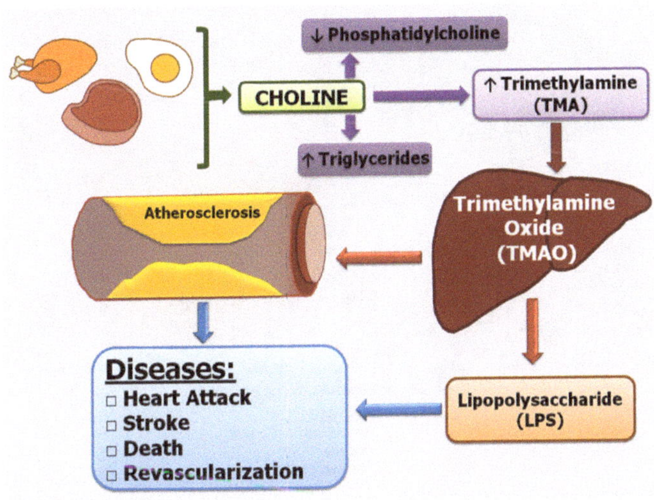

Fig. 3.9 Fatty liver and coronary diseases. (the microbial activity triggers the production of triglycerides and trimethylamine oxide)

If you harbour microbes that can metabolize TMA i.e. process it into the oxygenated variety known as trimethylamine oxide (TMAO), this can spell danger for your heart. The benign TMA falls prey to microbial manipulation in our gut that renders this relatively innocuous agent into a toxin with devastating impact on our well-being. TMAO is an important tell-tale sign of cardiovascular diseases. There are other intestinally derived toxins fuelled by microbial activities that can cause havoc to the normal functioning of our body. Sulphate derivatives of aromatic chemicals originating from microbial intervention in the intestine can be a major burden on the kidney and may lead to renal abnormalities. Controlling the microbes that are in the business of transforming harmless chemicals into noxious ones is critical strategy that need to be pursued in order not to succumb to the microbial imbalance promoted by fatty foods (Fig. 3.9).

3.8 What Mood Changes Have to Do with Gut-Dwelling Microbes

Even though we may not be completely aware there is an important communication corridor between the gut and the brain that is nourished by numerous signals emanating from the microbial activities in the intestine. The gut-brain axis relies on the messengers and/or their precursors generated by the microbes we harbour. They produce short-chain fatty acid (SCFA) like butyrate and neurotransmitters like γ-aminobutyric acid (GABA), dopamine and serotonin. Whenever you have a rewarding feeling due to some activity like eating that you are engaged in, blame

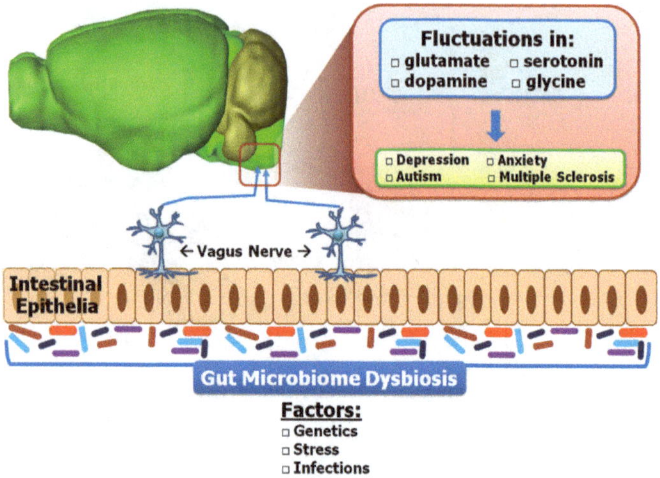

Fig. 3.10 Dysbiosis, abnormal metabolite production and neurological disorders

it on dopamine. These instruction-laden chemicals that are also produced by other organs in our body are central to numerous tasks executed by the brain and can stir our behaviour in one direction or the other depending on the concentration of a specific neurotransmitter. Hence, factors that modulate the flux of the microbial communities and their propensity to dish out these neuro-active commanders have tremendous influence on our health and well-being. For instance the lack of microbial diversity associated with ageing impedes the ability of the body to generate SCFA such as acetate, propionate and butyrate that supply the brain cells with energy. The loss and/or the rationing of the nutrients fuelling the brain result in a reduction of cognitive and memory power. This attribute is a common characteristic reminiscent of ageing. Hence, the propensity of people in retirement homes not to remember or retain mundane facts (Fig. 3.10).

The microbiome definitely has its imprint on the functioning of the brain, a feature that has yet to be fully deciphered. For instance, the amino-acid tryptophan is the precursor of numerous neurotransmitters that dictate a variety of our behavioural responses that are usually tailored to the stimuli we receive. This is an essential nutrient that humans cannot make with their own traditional organs and have to acquire it from the food we eat. However, depending on our microbial communities we can easily have ample supply as numerous gut microbes are capable of synthesizing this nutrient; but with caveat that the amounts of tryptophan we have access to need to be modulated if our response is to be commensurate to its presence or its derivatives. Any abnormal variation of this commodity in our body can become a harbinger of neurological complications. A shift in microbial communities induced by a myriad of factors can create havoc with blood tryptophan levels. This can then results in depression, mood swings and neurological ailments like Parkinson's or Alzheimer's diseases. Keeping a careful watch on our microbial partners and their ability to generate these brain responsive signals will lead to better health outcomes (Box 3.5).

Box 3.5 Microbial Communications with the Brain and Neurological Disorders

The ability of the microbiome to produce a variety of neuroactive chemicals makes the invisible organ a potent modulator of human behavior and an instigator of neurological diseases. For instance, Parkinson's disease is an incurable adult neurodegenerative disorder characterized by abnormal movements due to defective motor control. Dopamine production is impeded and there is an increased formation of aggregated neurotoxic protein known as α-synuclein. These two biochemical manifestations may be shaped by the nature of the microbial residents by limiting precursors of dopamine and activating the immune system with the release of pro-inflammatory chemicals like LPS. The gut microbiome in Parkinson's patients is usually characterized by a decrease in *Prevotella* spp and an increase in *Enterobacteria* spp. A diminution of mucin synthesis increases intestinal permeability and leads to an acute dysbiosis that contributes to the abdominal discomfort, bloating and premature satiety reported in these patients. Autism spectrum disorders (ASD) are another set of neurological perturbations leading to social and behavioural impairments including repetitive behaviours and non-standard communications. Here again, the microbiome may be a contributing factor. A decrease in *Veillonella* spp coupled with a rise in *Clostridium* spp and *Campylobacter* spp have been reported. Microbial products like p-cresol has been observed in elevated amounts in autistic children and 4-ethyl sulfate (4-EPS) has been shown to trigger anxiety like behavior. This microbial link is prompting the search of probiotics (e.g. *Bifidobacterium fragilis*) that may help in curing these diseases.

3.9 Microbiome and the Cancer Connection

Cancer is a multi-factorial disease that is a major global concern. Almost everyone knows of someone who has been affected by this disease. Numerous governments have recognized this disease as a national priority and have decided to provide all the support necessary to fight cancer. Although life-style, genetic disposition and environmental factors are important contributors to the spread of this disease, it is becoming amply clear that our microbial landscape and cancer proliferation are interconnected. It is being increasing recognized that the microbiome plays a key role in preventing and promoting cancer. The gut microbial ecology provides an important cover against invading pathogens that contribute to the onset of tumour formation. However, a shift in microbial communities induced by diet and pollution creates a favourable environment for opportunistic microbes to thrive and neutralise the protection offered by the finely tuned original microbial guards. The invading microbes produce virulence elements responsible for unimpeded cellular growth, a key feature of all cancers. They also put in motion processes that result in the instability of the DNA and furthering the progress of cancerous cells. Understanding the communal work of the intestinal microbes is pivotal if this problem is to be tackled. Indeed it has been shown that Africans from rural areas exhibit lower risk of colon and rectal cancers due to the increased abundance of the microbes known as *Bacteroides* spp. These are avid producers of short-chain fatty acid like butyrate, an important energy-producer and a promoter of cellular growth. This is a critical signaling molecule that can goad the body to synthesize sentinels that stop cancer cells on their tract.

On the other hand, the group of individuals whose colon is home to higher amounts of *Prevotella* spp tend to have higher levels of bile acid and lower amounts of butyrate. This switch in colonic microbial ecology is suspected to be responsible for the colorectal cancers (Fig. 3.11). The *Prevotella*-rich individuals are programmed to secrete bile acids in the feces. These compounds are known to promote the proliferation of cancerous cells. The interplay between these two microbial communities and their ability to produce these metabolites afforded wellness to one group and colorectal cancer to the other. Hence, the promotion and nurturing of functional microbial communities and limiting the presence of unregulated microbial ecosystem in the gut are pivotal to maintaining our well-being. This invisible organ has to be tended in the same manner like any other visible organ; one cannot expect to have working lungs supplying the furthest nooks in the body with oxygen if one indulges in smoking or toiling in occupation

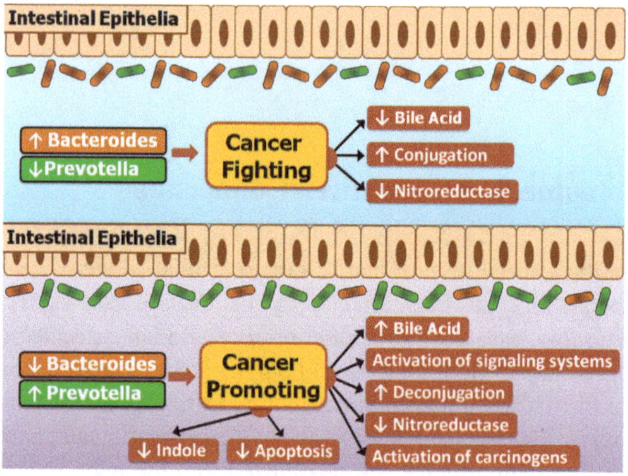

Fig. 3.11 Colorectal cancer triggered by microbial imbalance in the colon. (Decrease in detoxification and increase in cancerogenic metabolites and enzymes)

with toxic air. This cardinal rule also applies to the microbiome even if its anatomical features do not readily pop out to our eyes. Otherwise we run the risk of turning this organ that we almost never see but hear rumblings of once in a while into a cancer-causing entity. The critical attributes of these microbes need to be tapped as cancer preventers and fighters.

3.10 The Road to Our Well-being: The Microbiome Way

The microbiome is an essential component of humans if not of all multicellular organisms. This invisible component of the body that quite often makes itself heard with a rumbling sound is involved in a variety of biological tasks that are essential for the proper functioning of any human. Although these microbes residing in and on us have their genetic information they are bound by the communal existence they have adopted. The role and behaviour of each member are dictated by the host and other members of the collectivity. This harmony and synchronicity of purpose force everyone to comply with the rules set-up during the establishment of the community. However, these communities evolve in relation to the host and the changing landscape they have to deal with. They accompany us from the very beginning when we are developing till death. It is becoming more apparent that these microbes contribute to our development and eventually mould us the way we are both anatomically and physiologically. They have a say virtually in all aspects of our life and prod us as we go about our daily living. From providing essential nutrients like vitamin K that we are dependent on but incapable to make on our own to protecting us against opportunistic organisms and helping heal our wounds, these microbes partake in a myriad of tasks just as visible organs do. However, their constituents and interactions like all other organs in the body are modulated to large extent by the changes in environment, food, stress, and hormonal fluctuations we are subjected to.

The invisibility of this organ and its dispersed anatomy have contributed to the relegation of the microbiome to a low esteem in the hierarchy of our body parts. But this view is being discarded at a rapid pace as intriguing information about its role is being revealed. Its intimate link to our existence was only visualized very recently. What you cannot see, you cannot appreciate. This dictum fits perfectly the microbiome. The advent of next generation sequencing (NGS) coupled with bioinformatics tools is laying bare the indispensability of the microbial communities roaming within and on us to our very existence. Our life will not be the same without our intricate network of microbial communities. Hence, understanding how this invisible communal habitat is constituted, how it evolves and how the various partners communicate amongst themselves is crucial for our being. The host will not exist without the microbiome. It is as crucial as the heart and any other body part. This mission is as critical as our desire to unravel the workings of the brain or to venture in the outer galaxy. Its systematic functional identification will be akin to the discovery of a new organ as in occurred in the

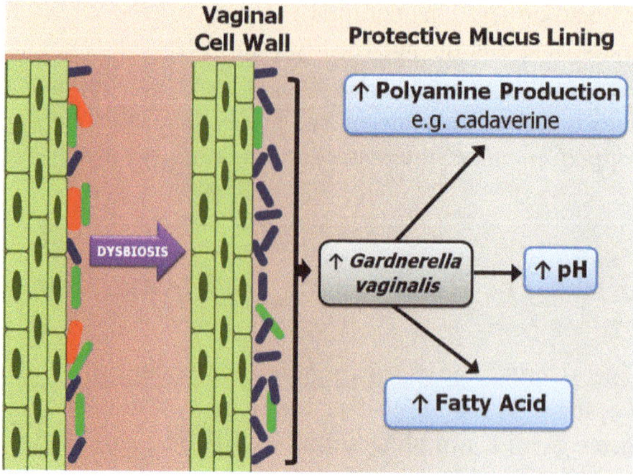

Fig. 3.12 Microbial proliferation and Vaginosis. (increased pH and polyamine smell are common markers of this disease)

sixteenth century when human was first beginning to comprehend the workings of various body parts. Just imagine the excitement when the Italian surgeon, Realdo Columbo identified how the four valves of the heart permitted the flow of blood in only one direction i.e. from the right ventricle to lungs and back to the left ventricle and on to the aorta. We are on the brink of tantalizing discovery on a part of our body that has long been ignored due to the lack of proper tools to decipher the microbial communities. It is indeed more exciting as this invisible organ has at least 1,000 times more genes than our own visible body. The microbiota we possess must play an important role in our development and we are physically and physiologically the way we are because of them. We would be entirely different if it was not for the microbiome and this holds true to most if not all organisms wandering on this planet and most probably beyond.

The proper functioning of this community with multiple partners depends on the finely tuned relationship among all the constituents. Every member operates in a mode that is beneficial to this cooperative life-style where interdependence is the modus operandi. However, this harmony can be perturbed by a variety interfering influences and can result in a multitude of ailments. The intrusion of opportunistic organisms fuelled by a change in the ecosystem gives rise to dysfunctional communities that are the precursors of debilitating diseases like cancers, IBS, and vaginosis. The latter is a disease arising from the uncontrolled proliferation of microbes such as *Gardnernella vaginalis*, *Veillonella spp*, and *Bacteriodes spp.* that promotes an increase in pH, fatty acids and polyamines (Fig. 3.12). Cadaverine, a polyamine secreted during this dysbiosis is the cause of the malodor characteristic of the disease. Once the fine balance involving the constituents and the signals that are responsible for a proper microbial community is uncovered, we will be in a better position to predict what makes us tick and what ails us. This knowledge is just now beginning to emerge and ignoring this important aspect of our body will be to our peril; it would be akin to not wanting to know how the brain functions (Fig. 3.12; Fig. 3.13).

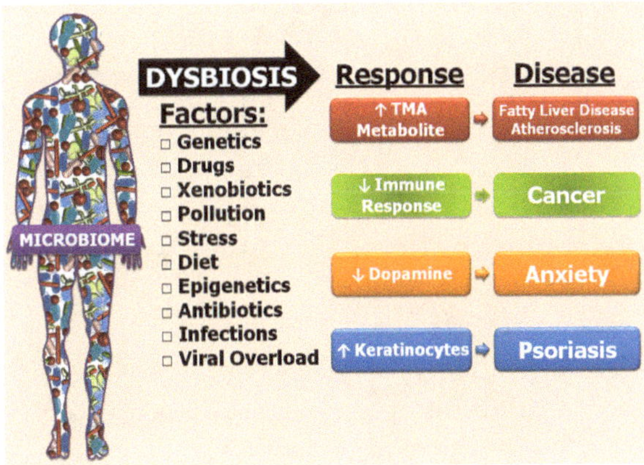

Fig. 3.13 Summary of dysbiosis: causes and diseases

3.11 Probiotics, Prebiotics and Synbiotics: The Nurturing of the Microbiome

The microbial communities that constitute our microbiome are dependent on a myriad of factors such as the genes we inherit, our mother's milk, the environment we live in, our hobbies, the seasonal changes we are subjected to, the food we eat, the life-style we lead and the medications we take. Despite the ability of our microbiota to respond and adjust to these situations, the invisible organ can be influenced by either taking in select beneficial microbes with known functional attributes or by consuming foods and plant products that promote the proliferation of specific microorganisms. This microbiome-rearing strategy is akin to feeding our brain with the books we read or the educational programs we watch or the games we play. Remember some eager mothers keen on giving their children a head-start in life, read to them or play music even when babies are in the womb. To promote a healthy life-style, seniors are being encouraged to partake in neuron-stimulating games on a regular basis. This brain exercise has becoming a common place in residences for the elderly. Thus probiotics, prebiotics and synbiotics can be utilized to guide the microbiome to perform optimally and ameliorate our well-being (Fig. 3.14).

Probiotics are live microbes administered as foods or even in capsules that confer health benefits on the body. Microbial cultures have been consumed by humans since the dawn of civilization. Mongolian women sprayed fermented milk on horsemen and their horses in belief that this provided strength and health on the recipients, while in some civilizations fermented milk was utilized to treat a variety of ailments. Dahi, a fermented milk is widely consumed in the Indian Sub-Continent and is known to impart numerous healthy outcomes. However, the first clinic trials on the health claims of these bacteria-rich fermented foods were not performed until the twentieth century. Eli Metchnikoff who obtained a Nobel laureate in medicine was first to correlate the longevity of some Bulgarian citizens

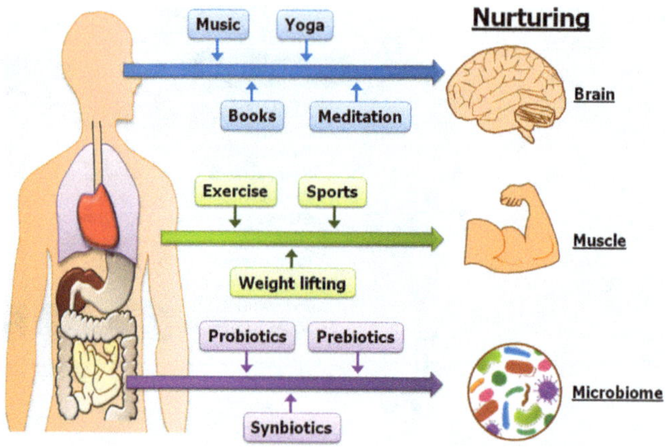

Fig. 3.14 Nurturing of the microbiome, the brain and the muscle. (activities and nutrients involved in fortifying these organs)

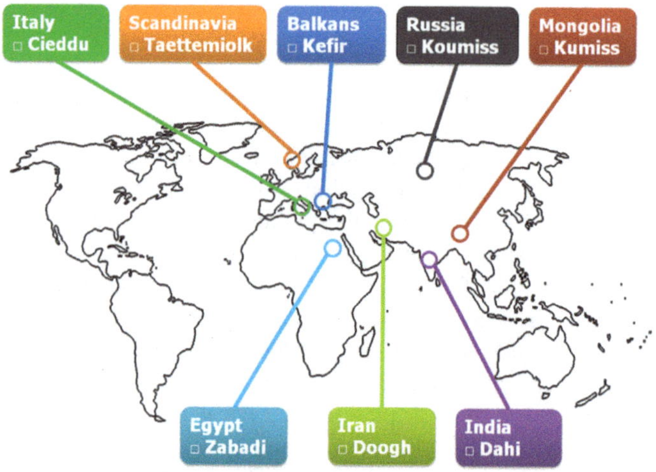

Fig. 3.15 Consumption of probiotic-rich foods around the globe

to the prolific consumption of fermented milk that was rich in *Lactobacillus bulgaricus* and *Streptococcus thermophilus*. In 1930, the Japanese scientist Shirota isolated microbes with probiotic properties from healthy human subjects. These were later utilized in the development of milk products that were commercialized as Yakult. Subsequently, a French pediatrician reported the lack of Bifidobacteria in stool of infants suffering from diarrhea. In 1965 the term probiotics was coined by Lilly and Stillwell to describe products that stimulate the growth of microorganisms beneficial to the body. It was in 1984 the first probiotic species, *Lactobacillus acidophilus* was introduced followed subsequently by *Bifidobacterium* spp. Their use in food products is widespread. The significance of probiotics in fortifying the immune system has officially been recognized by the World Health Organization (WHO). Currently a wide variety of microorganisms such as *E. coli, Propionobacterium, Enterococcus, Streptococcus, Leucomostoc*, and *Bacillus cereus* are being consumed in order to regulate and adjust the body's microbiome (Fig. 3.15, Box 3.6).

Box 3.6 Probiotics: Occurrence and Uses Worldwide
Humans have been consuming microbe-laden foods unknowingly since the dawn of civilization. Fermented milk, vegetables, meat and fish were and are a regular part of the daily diet for most people around the globe. Recently probiotics enriched foods like yogurt, yakult, kefir, dahi and cheese have become a regular repertoire of natural products that nutritionists are eager to extoll the health virtues of. *Lactobacillus* spp, *Bifidobacterium* spp and *Bacillus* spp. are the more prominent probiotics even though other microbes are being added to this list. They are also being used in non-consumable items like oral care gel and anti-ageing serum. Their ability to produce acid, enzymes, SCFA, immune responsive factors and to help establish functional microbial flora have made these probiotics excellent candidates to cure a variety of diseases that are provoked by dysbiosis. While *E.coli Nissle* provides relief to patients suffering from ulcerative colitis, a disease like vaginitis can be remedied with *Lactobacillus rhamnosus* GG. Irritable bowel syndrome victims can find comfort with the intake of *Bifidobacterium animalis* and *Lactobacillus acidophilus*. The role lactic-acid microbes like *Bifidobacterium breve* has been recognized in diminishing high blood pressure due to their ability to produce vitamin D, anti-hypertensive factors and interrupting cholesterol absorption. Although influence of probiotics in promoting wellness and as therapeutic agents aimed at numerous illnesses are becoming more prominent, their usage will become universal once the dosage and the intake frequency of these microbial supplements have been properly evaluated.

3.12 How Probiotics Work

Probiotics not only help the body to fine-tune the microbiome, they can also perform some very specific functions. Depending on the probiotics, these microbes act as a barrier and prevent colonization by opportunistic bacteria. They do so by re-enforcing the protective function of the gut mucosa and by generating signals that arrest the invasion by infectious organisms. They produce antibiotics, antioxidants and improve the body's immune system. Probiotics are also known for their ability to synthesize SCFA and vitamins, ingredients essential for numerous metabolic activities. They secrete enzymes that help in digestion and in the elimination of wastes. A good probiotic should be able to impart health benefits to the host and be bestowed with no pathogenic properties. Its presence in the microbial community should enhance some biological functions and/or curtail any negative influence on an individual's well-being. These can preferably be taken orally and must survive the harsh environment of the digestive system, especially the stomach where the pH can be unforgiving. The low pH of the stomach can be avoided if the probiotics are mixed with food such as milk, dietary fibres and yogurt. They can be designed to generate specific biomolecules that can be of benefit to the host. For example, lactose-intolerant individuals can consume probiotics with the ability to synthesize lactase or galactosidase that helps in metabolizing the milk sugar, lactose.

Probiotics have virtues almost like stem cells. Stem cells i.e. cells that have not yet made up their mind what they will become or which organ they will develop into can be utilized as a therapy to fortify ailing components of the body. For instance, they can be injected in the cornea or the liver with the proper commands that enable them to mature into corneal or hepatic cells that become part of the rejuvenated organs with optimal biological functions. Probiotics act in a similar manner in regard to our microbial communities. They can be consumed with the aim of enhancing a specific function like in the case of a select targeted fortification that is desired in the body. Individuals suffering from the inability to digest milk are readily relieved of their aversion to milk or milk products by the consumption of *Lactobacillus*. This microbe secretes lactase, an enzyme that can clip the milk sugar lactose into glucose and galactose which are then utilized by the body. Probiotics such as *Aspergillus oryzae* found in fermented foods like soy sauce, and sake secretes amylase and lipases that help in the digestion of starchy and oily meals. In instances where the microbiome has been infiltrated by opportunistic microorganisms that prevent the invisible organ from performing its regular task, an intake of the probiotics can shift the microbial balance toward a more fruitful one for the body. One case in point is the colonization of the dental space by *Streptococcus*, a microbe responsible for carving cavities on the tooth. Food enriched with various species of *Lactobacillus* can rectify this situation by creating an unfavourable situation for the invading bacteria to survive.

Among other various biological weapons they have in their armoury, *Lactobacillus* group of probiotics can generate an acidic environment. This situation is known to impede the growth of the occupying bacteria and arrest their assault on the teeth. Unlike the stem cells that are mainly involved in regenerating desired tissue or organ, probiotics can tilt the balance of a non-functional microbiome towards a functional one by quashing undesirable elements in the community. They can also be engaged in a targeted task designed to thwart a discomfort an individual is experiencing as in the case of intolerance to lactose or abnormal movement of the digestive system. World-wide people have been consuming fermented products laden with probiotics. These foods have evolved through centuries and each region has its own speciality of delicacies full of microbes. These probiotics have become part of the culinary culture and it will not be surprising if we learn that our palates have evolved accordingly. Yogurt, a diary product supplemented with microbes *Lactobacillus, Bifidobacteria* and *Lactococcus* is part of the grocery shelves in almost all countries. In India dahi, a fermented milk product

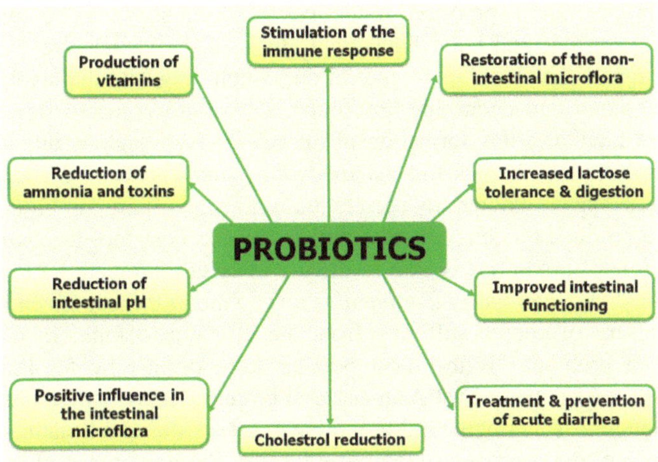

Fig. 3.16 Functions of probiotics in human health

rich in *Lactococcus lactis* and *Lactobacillus acidophilus* is taken with nearly all foods and its curative powers are readily touted. In the Caucus region, kefir with its bountiful of *Lactobacillus, Lactococcus*, *Leuconostoc* and *Bifidobacteria* is a daily staple. This is derived from goat's milk and contains yellowish grains resembling cauliflower. These grains are in fact polysaccharide capsules full of probiotics. In Japan the miso soup prepared from barley, rice, beans and rye is king while in Africa Uji the probiotics charged maize or sorghum or millet is widely cherished. Although not currently available probiotics with the ability to metabolize ethanol or cure obesity will be a welcome help to those individuals who are susceptible to adverse reaction upon intake of alcoholic drinks or have difficulty controlling their body mass index (Fig. 3.16).

3.13 How Probiotics Helps Adjust Our Microbial Community

The mouth is literally a continuation of our external environment and is being constantly challenged by the microbes we are exposed to in our surroundings. Even under this threat, the soft and hard components of the mouth harbour a good number of microbial communities that allow them to fulfill various essential tasks. The change in the chemical milieu in the mouth triggered by the intake of foods provides a fertile landscape for foreign organisms to set foot, proliferate and disturb the working microbial harmony. This happens at the onset of dental cavities when *Streptococcus* rapidly colonizes the tooth. In this instance, a probiotic like *Lactobacillus rhamnosus* can spring in action and create an acidic environment by the secretion of lactic acid. Its ability to produce antimicrobial factors administers another debilitating punch thwarting the *Streptococcus*-driven assault. This probiotic also helps train the immune system to fend the noxious microbe.

Bile, a product that is produced in the liver has been shown to be involved in a variety of tasks that make the body ticks. Its acid derivatives are emerging as important signals that allow us to control our weight, regulate the level of cholesterol in the blood and determine how much fat we are going to store. The key enzyme that facilitates this formation of the bile-derived acids is bile salt hydrolase. A number of probiotics like *Lactobacillus, Bifidobacteria, Bacteriodes and Enterococcus* spp is known to secrete this enzyme. These probiotics can be designed to produce this enzyme that will be an excellent tool to control weight gain. This role of probiotics is akin to the use of lactase-producing *Lactobacillus*. Intake of food enriched with *Lactobacillus* is a common therapy given in order to easy the misery of people suffering from the aversion to milk. In fact, lactase derived from probiotics is the most popular form of therapy for this ailment. Nearly 90–100% adults in East Asia cease to produce lactase as they reach adulthood. Although it is totally natural not to have lactase after the weaning years, the omnipresence of dairy products significantly raises the importance of this enzyme beyond our childhood. Hence, the remedy provided by *Lactobacillus*-supplemented foods or capsules is a boon to the adult population world-wide who wants to partake in their dairy-laden delicacies. The curing of enzyme deficient diseases or conditions with the aid of probiotics will become a common practice as our understanding of the microbiome becomes clearer (Fig. 3.17).

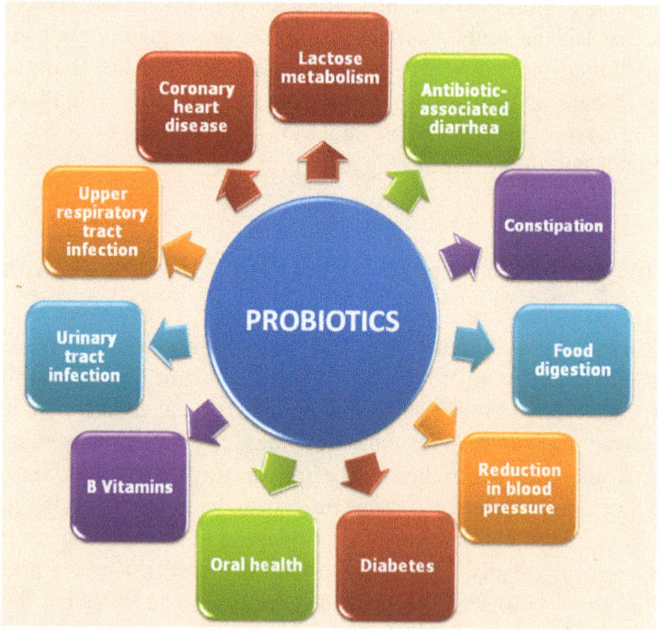

Fig. 3.17 Probiotics: the natural cure for diseases

In diabetic patients a microorganism referred to as *Akkermansia muciniphilia* that shelters in the mucus in the intestine can be an excellent candidate to promote carbohydrate metabolism. These microbes constitute 3–5% of the gut microflora and are known to stimulate glucagon – like peptide (GLP), an important modulator of blood sugar. Glucagon and insulin are the sugar police in the blood. Glucagon becomes active at night during asleep. This is when the blood sugar is low. The sugar sentinel ensures that we wake up healthy by maintaining the right sugar balance. Insulin on other the hand, protects against high blood sugar. For instance, when you indulge in a big bar of chocolate, it preserves the proper blood sugar and directs the extra to be stored. Hence, people with a hardy dose of *A. muciniphilia* are less insulin resistance, a major contributing factor to obesity. The abundance of these microbes decreases with age and the intake of high fat diet. The magic of *A.muciniphilia* is in the production copious amount of mucin, a slimy lining for intestine that promotes a diverse microbial ecosystem and fends off opportunistic microbes. Hence, the population of this blood-sugar balancing microbe can achieved by either taking it as a probiotic or by consuming nutritional fertilizer like fibers aimed at stimulating its proliferation. *Bifidobacteria* is another group of microorganisms that dissuade infectious microbes to colonize the gut by blocking the anchoring sites and stimulating the immune system. It is not surprising that they constitute almost 80% of the cultivable microbes in the stool of infants as opposed to the meagre 20% found in adults. They contribute to the defense power in babies as they are vulnerable at this tender age (Fig. 3.18a).

a

Microbial Markers	Diseases
Helicobacter pylori	Gastric ulcers
Propionibacterium acnes	Acne, eczema, allergy
Pseudomonas aeruginosa	Cystic fibrosis
Porphyromonas gingivalis	Gum disease
Enterobacterium spp.	Crohn's disease
Dialister spp.	Spondyloarthritis
Campylobacter spp.	Neurological diseases
Fusobacterium spp.	Appendicitis
Lactobacillus spp.	Vaginosis
E.coli	Colorectal cancer

Fig. 3.18a Microbes as biomarkers for diseases

3.14 Microbes at the Rescue of the Most Vulnerable: Infants and Seniors

Infants especially those who are born premature can also suffer from damaged intestinal tissues that begin to die. This disease known as necrotizing enterocolitis is characterized by bloating and swollen abdomen responds favourably to probiotics. Use of the probiotic *Lactobacillus* is quite effective in mitigating this disease. On the other hand, the intervention with a concoction of *Bifidobacterium, Propionobacterium* and *Lactobacillus* significantly reduces the risk of allergy in infants delivered by C-section. In fact these C-section babies are devoid of the microbes that mothers impart on them if they are not born naturally. Recent studies involving the sprucing the babies with the mums' microbiota are showing positive results. This kind of probiotic treatment where someone else or one's own microbe is used as probiotics to rectify aberrant biological activities is becoming more prevalent. Transfer of microbes from fecal matter, ear wax, skin and other body parts is gaining medical traction and are being implemented in health centers world-wide. Microbial transplant undoubtedly impart numerous health benefits (Fig. 3.18b; 3.18c).

Elderly individuals are another segment of the population who can benefit tremendously from the enrichment of their microbiome with probiotics. Ageing is characterized with a sharp decline in microbial diversity and density. This phenomenon is the leading cause of the decrease of metabolic activities and in some cases of physiological abnormalities observed in seniors. The lower abundance of *Bifidobacteria* coupled with the increased presence of *Enterobacteriaceae* is a common predictor of the ageing process and such a shift in microbial population is responsible for the diminished production of enzymes and essential vitamins like vitamin B that are critical in maintaining a constant supply of energy in the

b

Therapy: Microbes Suggested	Target Diseases
Lactobacillus; Bifidobacteria	Diarrhea
Lactobacillus salivarius	Gastric ulcers
E. coli 83972	Urinary tract infection
Salmonella typhimurium	Pancreatic cancers
Streptococcus salivarius	Halitosis
Staphylococcus epidermidis	Atopic dermatitis
Akkermansia muciniphila	Obesity
E.coli Nissle 1917	Ulcerative colitis

Fig. 3.18b Microbial therapy: Microbes to cure diseases

c

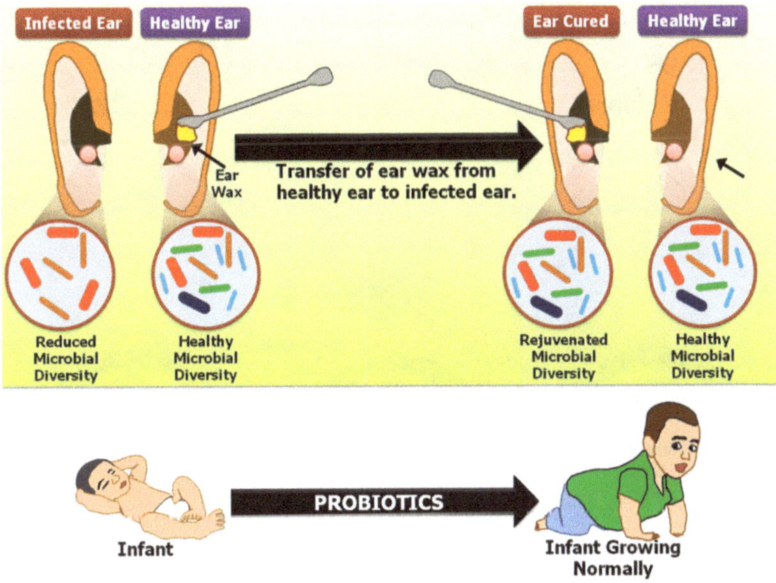

Fig. 3.18c Ear wax transplant and infant needs probiotics to develop

body. There is also a build-up of toxic products as the proper microbial communities involved in their decomposition are severely hampered. This situation is analogous to someone not going to the toilet on regular basis. The build-up of noxious elements in the colon can be the cause of a variety of ailments including head aches and digestive discomfort. The advanced age of centenarians is attributed to a diversity of microbial populations that are a hallmark of all individuals living beyond the ripe of age 100 years. The mission of these microbes may be compared to the role stem cells play in rejuvenating the organs and extending their functional life span (Fig 3.19). In this instance the diversity of the microbes observed in centenarians may be allowing these microbiota to supplement some of the functions their visible organs cannot perform or to generate chemical ingredients triggering the proper physiological responses from these ageing organs. Hence, harnessing the power of probiotics in reconfiguring the dysfunctional microbiome characterized by the ageing process can be an important game-changer in the life-style of seniors. This will be critical in seniors who are usually on microbiome-distressing medications. That is why it should not come as a surprise if elderly individuals take more time to heal and are invaded at a rapid rate by opportunistic bacteria. In fact microbial infection is a major concern when elderly patients are admitted to health facilities. They may be in for a hip replacement but unfortunately they run a higher risk of being contaminated by hospital dwelling-microbes. The prevalence of hard to eliminate microbes like *Clostridium difficile* in elderly patients is a problem world-wide. And often infectious outbreaks occur in these places. Unbalanced nutrition and erratic eating habits

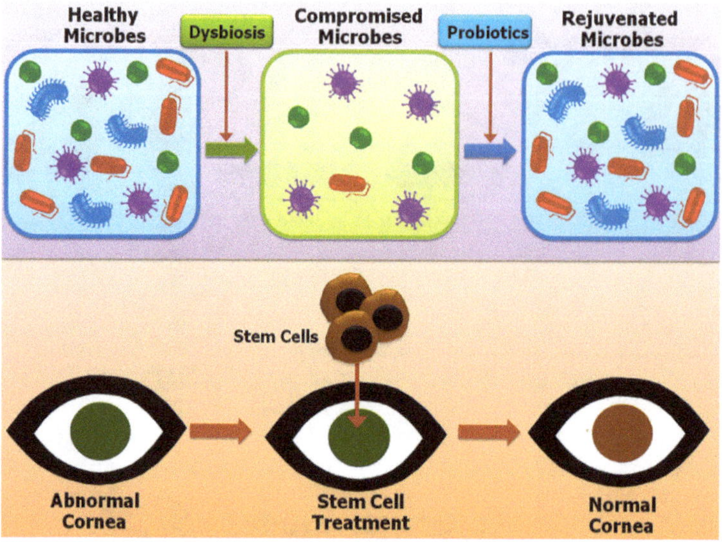

Fig. 3.19 Some probiotics are like stem cells and help in the rejuvenation of the microbiome. Stem cells in corneal treatment.

induced by reduced sensation in the olfactory and taste systems add to the dilemma facing seniors. These conditions create a nutritional environment deficient in essential nutrients like calcium and vitamins that further exacerbate the microbial ecosystem. Intake of the probiotic *Bifidobacterium longum* stimulates the proliferation of other *Bifidobacterium* spp. Administration of *Lactobacillus acidophilus* is known to increase the synthesis of antioxidants like glutathione and oxidant bursting enzymes like catalase. While the regular consumption of *Bifidobacterium animalis* helps reduce the time food transits in the gut, the probiotics *Lactobacillus rhamnonu*s and a *Propionibacterium* spp contribute to the improved defecation frequency. Hence the implementation of a nutritional regime rich in probiotics will go a long way in easing the discomfort and some of the ailments that are associated with old age.

3.15 Probiotics: The Disease Fighters

Evidence-based research is revealing how probiotics can help reverse ailments caused by the disruption of the microbial communities in our body. Irritable bowel syndrome (IBS) is one such disease where a perturbation in microbial population is the main cause of the disease. This shift in the ecosystem of the microbiota is punctuated by a loss in immune tolerance and a rise in inflammatory response. To mitigate this situation and to remedy this ailment, the administration of probiotics like *E.coli Nissle* has been relatively more effective than the use of antibiotics.

In this instance, the probiotics appear to deliver a three-pronged attack on the invasive and crafty microbes. They improve the fence lining the intestine, they out-compete the bad bacteria and they secrete molecular soldiers like hydrogen peroxide and lactate that stop invaders right on their tract. The significance of probiotics as therapeutic agents is only now beginning to be appreciated and is slowly challenging the aggressive use of antibiotics in treating these conditions. Unfortunately, antibiotic-based remedies tend to seriously interfere with the workings of our microbiome as their action against toxic microbes also eliminates some of our trusted invisible partners. The uncontrolled proliferation of the stomach residing *Helicobacter pylori* is a cause of concern as it is a risk factor for gastroduodenal ulcers and lymphomas. The intake of probiotics such as *Lactobacillus* and *Streptococcus thermophilus* has been demonstrated to correct this imbalance. The ability of these probiotics to secrete SCFA like acetate and butyrate is known to be a key factor in impeding the march of the over-reactive *H. pylori*. Wound healing and cancer prevention are other two activities that probiotics are also involved in. *Lactobacillus acidophilus* that is an excellent candidate with the power to heal injuries is known to arrest the colonization driven by the scheming *Pseudomonas aeruginosa*. The acidic environment nurtured by the *Lactobacillus* renders the ability to conquer the tear on the skin almost impossible. The propensity of probiotics to bind to mutagenic compounds and inhibit the production of carcinogens by opportunistic pathogens is central for their anti-tumour activity. The lactic acid producing bacteria such as *Lactobacillus acidophilus*, *Bifidobacterium bifidum* and *Streptococcus lactis* are routinely being recommended for the prevention of cancers. Some select tumours like in colorectal cancers are being treated with the aid of these life-microbial concoctions. Hence it comes as no surprise that the use of probiotics as a prevention of diseases and a therapy against some specific disorders is on the rise. The commercial value of this industry is into hundreds of billions of dollars, a figure that is further expected to increase as probiotics are safe and well-tolerated in children, premature infants and in the general population as a whole. Furthermore, unlike pharmaceutical products, probiotics are very easily administered mostly as components of foods and the lingering effects of these chemicals are quickly becoming a relic of the past.

3.16 Prebiotics: The Microbiome Fertilizers

The microbiome can also be nurtured and programmed with the assistance of prebiotics. These are natural products that foster an environment propelling the proliferation of some desirable microbes. In the same manner as we are always eager to give our visible organs a boost when they are not working well or when they are sluggish, we can revitalize our invisible organ. Our more traditional visible organs can be invigorated by taking supplements like vitamins or minerals that provide ammunition to various metabolic activities in the muscle, brain, heart or liver. For instance, athletes take creatine, a product responsible for energy stabilization to

Fig. 3.20 Prebiotics: the microbiome fertilizers

Prebiotic Active Ingredient	FOOD
Arabinogalactan	carrots, tomatoes
Pectin Oligosaccharides	apples, pears
Fructooligosaccharides	asparagus, bananas
Galactooligosaccharides	lentils, chickpeas
Xylooligosaccharides	oats, barley
Beta-glucan	mushrooms, seaweeds
Polyphenols	blueberries, prunes
Milk oligosaccharides	milk

improve their performance. Creatine and creatine kinase duo enables us to engage in physical activities quickly and fast. Without the intervention of this dynamic couple, daily chores will happen in a relatively sluggish manner. Individuals with excess body mass consume carnitine with the hope of inducing the various body parts in burning fats. Well, the constituents of the invisible organ can be strengthened with the intake of prebiotics.

These are like fertilizers that create the proper environment for the desired microbial ecosystem to flourish. They tilt the microbial community in a manner favouring some specific tasks to be performed. For instance, a prebiotic promoting the growth of lactase-secreting microbes will help lactose intolerant people while a prebiotic inhibiting the competitors of vitamin K-producing bacteria will be of immense assistance to patients having difficulty with blood coagulation. Prebiotics are excellent nutritional supplements involved in modulating and adjusting the microbiome that can undergo disruption due to ageing, infection, climate change and a sleuth of other factors one may be subjected to as one goes through one's daily activities (Fig. 3.20).

What are these magical prebiotics? They are usually plant derived fibers and complex carbohydrates contained in milk that help promote the growth of specific microbial ecology in our body especially in our digestive system. They are mostly resistant to digestion by the enzymes produced by the gut. They are metabolized only by our invisible organ and/or promote proper microbial communities that impart health benefits to the host. The production of short-chain fatty acids (SCFA-acetate, butyrate) in response to the intake of prebiotics can contribute to the energy budget of the muscle, can regulate cholesterol biosynthesis and can provide fuel to the colon. Stimulation of calcium absorption, reduction in infections and repression of allergic symptoms are some of the other activities prebiotic intake is known confer on the host. These physiological processes are of course a major boon to our well-being as these are like lubricants oiling our body machines. The prebiotics prime the microbiome and this invisible organ is reflective of the prebiotics consumed. As a toned muscle system is indicative of the time spent at the gymnasium, specificity and diversity of the invisible organ is a tell-tale sign of our dietary fiber-eating habit (Fig. 3.21).

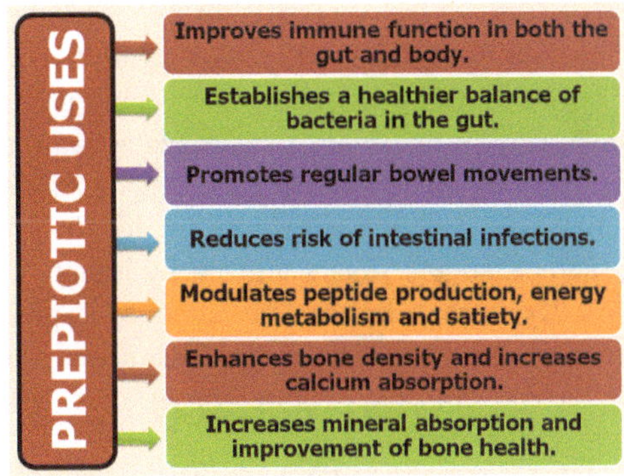

Fig. 3.21 Prebiotics and their health impact

3.17 Prebiotics in Everyday Foods

Prebiotics or dietary fibers have been part of human nutrition since the dawn of civilization. In 450 BC Hippocrates had noted the laxative attribute of coarse wheat compared to the refined variety. In the 1920s Kellogg sang the praise of bran as a nutrient that increased stool weight, prevented diseases and acted as a laxative. All health gurus extoll the praise of the nutritional quality of fiber-rich foods. After a relative quiet epoch in the health benefits of the dietary fibers, their involvement in mitigating ailments like obesity, cardiovascular disorders and diabetes were widely revived in the 1970s and has now become a staple of balanced nutrition. The main prebiotics are the galacto-oligosaccharides (GOS) obtained from milk, inulin associated-fructo-sugar (FOS) found in chicory roots and xylooligosaccharides (XOS) present in plant products like palm oil and corn plants. They also occur naturally in milk, asparagus, garlic, onion, leeks, wheat, oats and soya beans. They are soluble, resistant to the acidic environment in the stomach, fermentable and stimulate growth or the activity of intestinal microbes responsible for the well-being of the host. Prebiotics tend to be relatively selective to microorganisms like *Bifidobacteria* and *Bacteroides*. They are converted into simpler derivatives before they reach the colon. Here they act as an anchor for the establishment of a unique ecosystem that is aimed at aiding the host in accomplishing a variety of tasks. They generate energy that fuels the muscles and the colon. They are also involved in repressing allergic symptoms. The reduction of infection and promoting the proliferation of good microbes like *Akkermansia* are some other positive features prebiotics contribute to. They are also known to decrease glucose absorption, activate cholesterol excretion, and promote laxation.

One of the most intriguing influence of prebiotics in dictating our microbial ecosystem and promoting our well-being has been observed in infants just immediately after birth. Babies who are fed their mothers' milk have much better health outcomes than those fed formulated milk. They are less susceptible to allergic reactions and kids nourished by natural milk tend to be less prone to asthma. The GOS, a natural ingredient in the mother's milk cannot be digested by the babies' developing guts. They are designed to promote the growth of microbes that are crucial for a variety of tasks at this critical juncture in their lives. One can safely conclude that mother-nature is making these goodies not for the babies but for the microbes feeding on them. In fact, the presence of a unique ingredient known as sialic acid in the milk of some Malawian mothers result in well-nourished infants as opposed to infants who are fed by mothers without this magical tonic in the milk. This prebiotic is responsible for the growth of a unique set of microbes that generates simple sugars utilized by infants for their development. Even from such an early stage of human life, the body is programmed to depend on these invisible partners. It is not surprising to learn that various milk products are being formulated with oligosaccaharides in an effort to generate increased microbial population in growing infants.

The elderly can also benefit from the intake of prebiotics as their microbiota tend to undergo a drastic change and there is a significant reduction in the diversity and abundance of the microbial ecosystem as age progresses. The fructo-oligosaccharide (FOS)-based prebiotics have an excellent bifidogenic activity while resistant starch like sprouted legumes improves bowel movement by promoting the growth of *Lactobaccilli* and *Bifidobacteria*. These microbial fertilizers can be given to relieve seniors of constipation, a condition quite acute in this segment of the population. Prebiotics not only promote a healthy life, their intake can be tailored to remedy a number of ailments that are due to dysbiosis i.e the perturbation of microbiota in the body. Their ability to nurture a select group of bacteria that possess a range of tools to combat the opportunistic invisible invaders can be put to good use.

Obesity is a disease that is characterized by low grade inflammation and is punctuated by a decrease in the diversity of gut microbiome, a situation resulting in the ineffective harvesting of calories from what we eat. The inadequate energy extraction from food intake automatically triggers the hunger hormone, ghrelin which stimulates the desire to eat more food. This vicious circle causes the body to gain weight. The utilization of FOS prebiotics like inulin-laced pasta decreases the bacterial LPS, suppresses the hunger signal, promotes satiety and increases microbial diversity events aimed at countering obesity. A prebiotics is an effective means in controlling weight gain. Cardiovascular diseases do also respond well to prebiotics therapy. Water soluble fibers such as pectin found in fruits and guar gum from guar beans are known to decrease the bad cholesterol (LDL low density lipoprotein) without affecting the good cholesterol (HDL high density lipoprotein). They also help lower the blood pressure. The positive influence of these prebiotics on the functioning of the heart has been widely documented and accepted as a means of averting cardiovascular diseases. Mitigating constipation and prompting proper bowel mobility are the gold standard treatment fuelled by the intake of prebiotics. Cereal fibers like wheat bran are religiously utilized to counter any abnormality associated with fecal bulking. Prebiotics can also be a potent remedy against cancers. Enhanced micronutrient absorption and stimulation of microbes with cancer fighting enzymes like glutathione transferase are some of the mechanisms that these nutritional elements confer to the body. The ability of prebiotics to enrich the diversity of the microbiome is central to its curative power (Fig. 3.22; Fig. 3.23; Fig. 3.24).

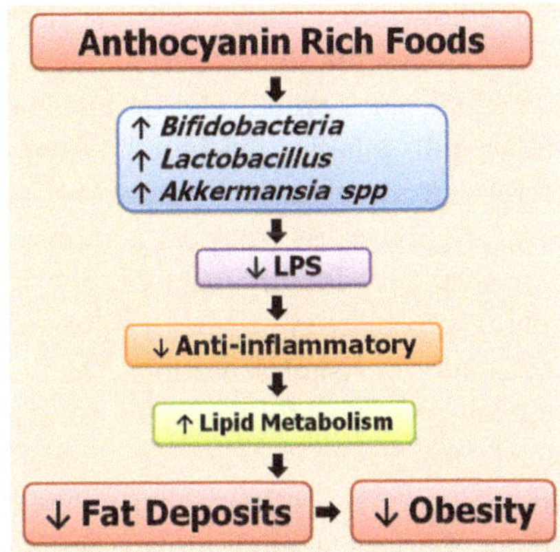

Fig. 3.22 Prebiotics: Anthocyanins against obesity

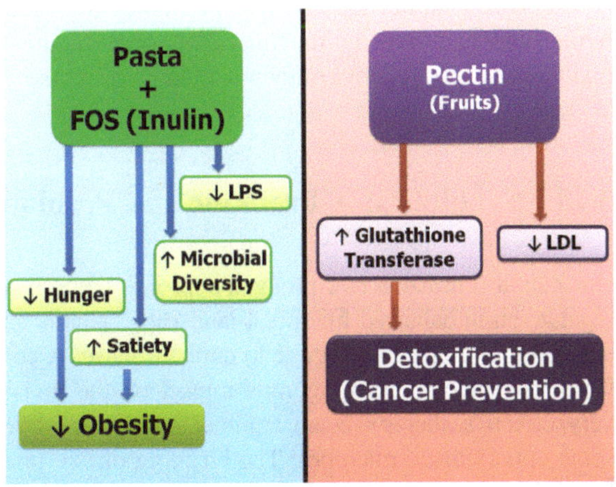

Fig. 3.23 Prebiotics and their molecular interactions

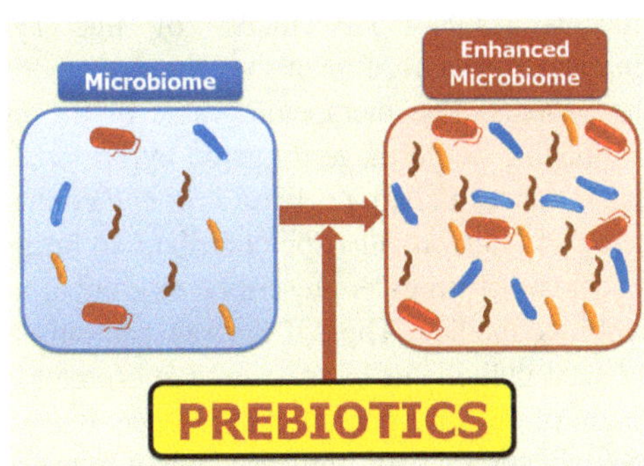

Fig. 3.24 Prebiotics therapy

Prebiotics is almost like a dietary manure that transforms the microbial landscape in the gut. A similar strategy is utilized to help the visible body parts if something goes amiss. During anemia when the level of hemoglobin is low, iron sulfate pills enables the blood system to stimulate the production of this oxygen carrier. In the case of keratoconus, an eye condition where the cornea loses its focusing power, the surgical administration of vitamin B2 allows the eye to function normally. In this instance, the infusion of vitamin B2 promotes the cross-linking of collagen, a biological event that fortifies the cornea and enables it to contribute to visual sensation. The same can be said of lipoic acid and calcium supplements. The former helps fortify the workings of the brain while the latter enable the proper functioning of the skeletal system. As these ingredients assist the target organs optimize their biological roles, the intake of prebiotics provides a proper nutritional environment for the right microbial ecosystem to flourish in order to maximize the output of the invisible organ in promoting a healthy body. However, this is a non-invasive exercise and usually corresponds to an accompaniment of the food we consume unlike the surgical intervention needed to feed the cornea with vitamin B2. Hence, prebiotics are natural fertilizers for the microbiome and can remedy the microbial perturbations that the body has to often undergo due to age, environmental change or intake of medications.

3.18 Synbiotics: A Probiotic and Prebiotic Concoction

One can also utilize a two-pronged approach comprising probiotics and prebiotics to regulate an unbalanced microbial landscape. This is termed as synbiotic and can be equated to a multi-drug regime to cure a disease. A combination of these two stimulators can have a more vigorous impact on the invisible organ and can provide a very effective therapy against ailments mediated by uncontrolled proliferation of some opportunistic microbes. The blending of live microbes and the microbial fertilizers that enable them to thrive is a very powerful tool to goad the constituents of the invisible organ in executing a dedicated task. For instance, muscle wasting and inflammation that characterize cancer cachexia has been shown to be mitigated with a concoction of a synbiotic consisting of *Lactobacillus reuteri* and inulin-based fructans (FOS). The intake of this synbiotic reduces the population of *Enterobacteriacae*, replenishes the *Lactobacillus* colonies and prolongs the life of the patients. The therapeutic value of the FOS and *lactobacilli* taken together in combatting hemorrhage triggered by bacterial infection has also been recognized. In this instance, a sharp decrease in *E.coli* counts coupled with an increase in the population of *Lactobacillus* spp contribute to bringing relief to these patients. Quite often a cocktail of drugs is prescribed to combat a viral infection; this can be equated to synbiotic therapy. The AIDS virus, can only be eliminated from the blood system of infected individuals by the use of anti-proteases and medications inhibiting the replication of the viral genetic code. Hence, this double punch aimed at aiding the microbiome naturally is of immense benefit to the body.

3.19 Conclusion

The human body is made of cells that are in constant flux due a variety of factors including wear and tear as age progresses. These can be repaired or rejuvenated with the aid of medical intervention. One can seek the help of a cosmetic surgeon to remove ageing spots or modify the shape of the nose. In case of the microbiome, any perturbations (dysbiosis) that the invisible organ experiences can be modulated by the programmed intake of probiotics and prebiotics. Hence, the microbiome behaves in manner that is comparable to the organs composed of visible cells. It is clear that the body is made up of visible and invisible cells. Most visible cells assemble as various organs we can see; there are few exceptions like the blood system where the cells are mobile and interact directly and indirectly with all the organs. Just like dysbiosis that robs the microbiome of its natural functions, the organs can also be subjected to a number of challenges over body's life span. The inherent repair machinery cannot remedy the situations the organs are facing in a timely manner, intervention in form of medication or surgery is needed to bring them back as contributing members of the body. For example if some oily clutters are impeding the flow of blood, the arteries are unclogged by surgical procedure. If organs of the body are insulted by overuse of some products or by accidental intake of toxic compounds, antidotes are utilized to bring back these anatomical structures in tip-top shape. For instance, fomepizole can give reprieve to a liver challenged by alcohol abuse while high levels of cholesterol in the blood that contribute to plaque build-up in the arteries can be rectified by statin family of cholesterol-busters (Fig. 3.25).

In the same manner any disturbance in the microbial communities constituting the microbiome can be adjusted with the intake of probiotics, prebiotics or

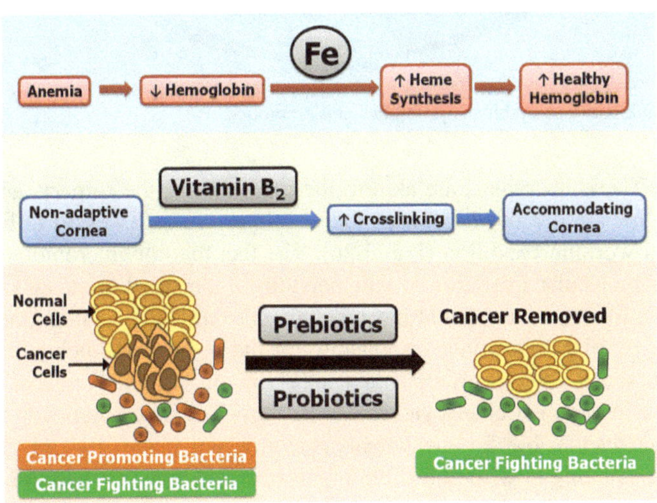

Fig. 3.25 Rejuvenation of visible and invisible organs

Fig. 3.26 Probiotics and prebiotics to counter dysbiosis and diseases

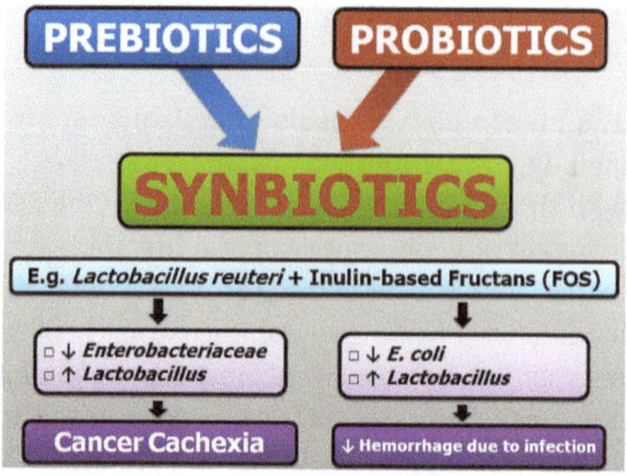

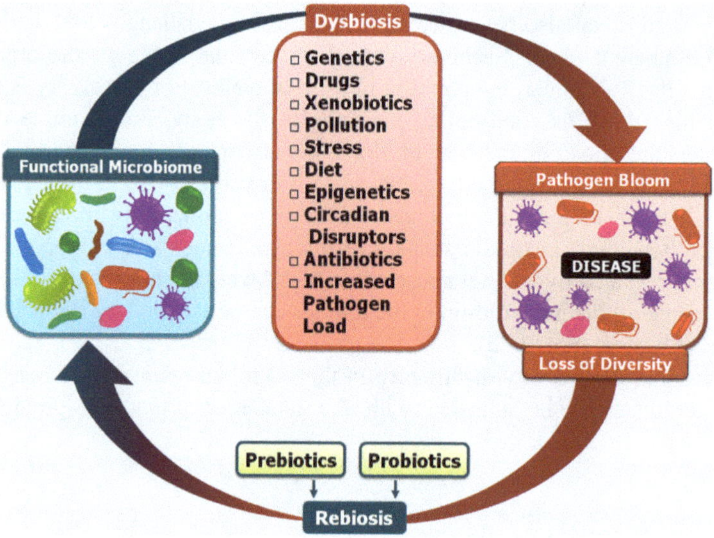

Fig. 3.27 Link between dysbiosis, probiotics and prebiotics

synbiotics. These 'therapies' are akin to the medications, the surgery, antidotes or stem cells that are deployed to tweak the visible organs in an effort to bring them to a normal working condition (Fig. 3.26). The fact that the microbiome is invisible and its molecular operation is still not fully understood compared its visible counterparts like the liver should not preclude it from being an important component of the cellular collectivity responsible for the body's well-being. The microbial system spans a wide surface area within and on the body just like the blood system as it imparts extremely valuable functions. This microbial ecosystem is to some extent responsible for the human anatomy and makes it work the way it does. Understanding how the invisible organ operates, how the harmony amongst its cells are disrupted and how the fine-balance is reconstituted with the aid of probiotics and prebiotics are essential if we are to understand the intimate details about ourselves. Ignoring these invisible collaborators is like disregarding the majority our own cellular components. Hence, dysbiosis and rebiosis propelled by the infusion of live seeds (probiotics) and the fertilizers (prebiotics) will go a long way toward promoting wellness and will bestow upon us a healthy life (Fig. 3.27).

Suggested Readings

Ambalam P, Raman M, Purama RK, Doble M (2016) Probiotics, prebiotics and colorectal cancer prevention. Best Pract Res Clin Gastroenterol 30:119–131. https://doi.org/10.1016/j.bpg.2016.02.009

Ashraf R, Shah NP (2014) Immune system stimulation by probiotic microorganisms. Crit Rev Food Sci Nutr 54:938–956. https://doi.org/10.1080/10408398.2011.619671

Baumler AJ, Sperandio V (2016) Interactions between the microbiota and pathogenic bacteria in the gut. Nature 535:85–93. https://doi.org/10.1038/nature18849

Bolnick DI, Snowberg LK, Hirsch PE et al. (2014) Individual diet has sex-dependent effects on vertebrate gut microbiota. Nat Commun 5:4500. https://doi.org/10.1038/ncomms5500

Brussow H (2013) Microbiota and healthy ageing: observational and nutritional intervention studies. Microb Biotechnol 6:326–334. https://doi.org/10.1111/1751-7915.12048

Ciorba MA (2012) A gastroenterologist's guide to probiotics. Clin Gastroenterol Hepatol 10:960–968. https://doi.org/10.1016/j.cgh.2012.03.024

Clemente JC, Ursell LK, Parfrey LW, Knight R (2012) The impact of the gut microbiota on human health: an integrative view. Cell 148:1258–1270. https://doi.org/10.1016/j.cell.2012.01.035

Culligan EP, Hill C, Sleator RD (2009) Probiotics and gastrointestinal disease: successes, problems and future prospects. Gut Pathog 1:19. https://doi.org/10.1186/1757-4749-1-19

Derikx LA, Dieleman LA, Hoentjen F (2016) Probiotics and prebiotics in ulcerative colitis. Best Pract Res Clin Gastroenterol 30:55–71. https://doi.org/10.1016/j.bpg.2016.02.005

Gilbert JA, Quinn RA, Debelius J et al. (2016) Microbiome-wide association studies link dynamic microbial consortia to disease. Nature 535:94–103. https://doi.org/10.1038/nature18850

Hutkins RW, Krumbeck JA, Bindels LB et al. (2016) Prebiotics: why definitions matter. Curr Opin Biotechnol 37:1–7. https://doi.org/10.1016/j.copbio.2015.09.001

Kilian M, Chapple IL, Hannig M et al. (2016) The oral microbiome - an update for oral healthcare professionals. Br Dent J 221:657–666. https://doi.org/10.1038/sj.bdj.2016.865

Kobyliak N, Conte C, Cammarota G et al. (2016) Probiotics in prevention and treatment of obesity: a critical view. Nutr Metab (Lond) 13:14. https://doi.org/10.1186/s12986-016-0067-0

Mileti E, Matteoli G, Iliev ID, Rescigno M (2009) Comparison of the immunomodulatory properties of three probiotic strains of Lactobacilli using complex culture systems: prediction for in vivo efficacy. PLoS One 4:e7056. https://doi.org/10.1371/journal.pone.0007056

Noval Rivas M, Crother TR, Arditi M (2016) The microbiome in asthma. Curr Opin Pediatr 28:764–771. https://doi.org/10.1097/MOP.0000000000000419

Patra V, Byrne SN, Wolf P (2016) The skin microbiome: is it affected by UV-induced immune suppression? Front Microbiol 7:1235. https://doi.org/10.3389/fmicb.2016.01235

Penney NC, Kinross J, Newton RC, Purkayastha S (2015) The role of bile acids in reducing the metabolic complications of obesity after bariatric surgery: a systematic review. Int J Obes (Lond) 39:1565–1574. https://doi.org/10.1038/ijo.2015.115

Rastall RA, Gibson GR (2015) Recent developments in prebiotics to selectively impact beneficial microbes and promote intestinal health. Curr Opin Biotechnol 32:42–46. https://doi.org/10.1016/j.copbio.2014.11.002

Rook GA (2013) Regulation of the immune system by biodiversity from the natural environment: an ecosystem service essential to health. Proc Natl Acad Sci U S A 110:18360–18367. https://doi.org/10.1073/pnas.1313731110

Sanford JA, Gallo RL (2013) Functions of the skin microbiota in health and disease. Semin Immunol 25:370–377. https://doi.org/10.1016/j.smim.2013.09.005

Sarkar A, Lehto SM, Harty S et al. (2016) Psychobiotics and the manipulation of bacteria-gut-brain signals. Trends Neurosci 39:763–781. https://doi.org/10.1016/j.tins.2016.09.002

Sheflin AM, Whitney AK, Weir TL (2014) Cancer-promoting effects of microbial dysbiosis. Curr Oncol Rep 16:406. https://doi.org/10.1007/s11912-014-0406-0

Chapter 4
What If the Workings of the Microbiome Are fully Uncovered? – A Revolution in Human Health, Wellness and Beyond

Contents

Abstract The world will be an entirely transformed place if in not too distant future the molecular workings of the microbiome are unlocked. Just imagine that the biogenesis of the microbiome is known, the regulatory processes mediating the assembly, disassembly and re-assembly of the invisible organ are unravelled, the communication networks among the microbes and the host are resolved, the genetic information and the role of each resident in the human microbial community are deciphered, and the interactions between the invisible organ and the visible organs are decoded. Such an accomplishment will be similar to the current state of knowledge we have on a number visible organs including the liver or the heart. This information will be revolutionary as it will have a tremendous impact

V.D. Appanna, *Human Microbes - The Power Within*,
https://doi.org/10.1007/978-981-10-7684-8_4

on the health delivery system where the microbiome will be central in promoting wellness and fighting diseases. Tracking of individuals and their behaviours with the aid of their microbial partners and the corresponding molecular fingerprinting will be a common practice in all aspects of human activities in a manner analogous to the currently widely utilized DNA profiling. Material belongings can be infused with characteristic personal microbes and identified readily. Commercialization of microbial communities endowed with a desired attribute will be a reality either as medications, cosmetics or to impart unique features to any individual. Designer babies, tailored-made non-invasive therapy for various diseases and a microbiome-based security system will be within the realm of possibilities. The microbiome-driven engineering of unique trait in any individual can also be performed on demand. The invisible organ has a lot of amazing features in store that will enable limitless societal transformation. The exploration of this treasure trove has just begun and it will undoubtedly reveal information that will reshape all spheres of our lives.

Keywords Designer babies · natural cures · diagnostics · forensics · physiological traits on demand · outer space living.

4.1 Gazing into the Crystal Ball – The Future

What if we knew everything about the invisible organ? Or even as much as we are presently aware of the visible organs say the liver – how our lives will change? How this knowledge can be utilized to improve our health and the world around us? The current understanding of the human microbiome is in its infancy. The degree of comprehension of this marvellous partnership we have struck with the invisible world that live within and on us can be equated to the same level of know-how the Dutch naturalist, Van Leeuwenhoek had on microbes following his invention of the microscope. He was happy just seeing and cataloguing the occupants of the invisible world where ever they happened to be. In the rain, on his teeth and on pets anywhere and everywhere he could find them. Scientists in that era were just eager to go on the hunt for places these minute creatures were in and it was only after many centuries the isolation, culturing and characterization of microbes were undertaken. Their functions, their significance in whatever environment they occupy and their utilization for the good of humankind came much later and is still being pursued to this date. The recent advent of DNA technology and next generation sequencing (NPS) have allowed scientists to peer through the invisible world in a manner that has never been done before. Even those microbes which do not survive when dislodged from their communities can be profiled. These culture-independent analytical tools provide a unique window on the microbial communities wherever they are in a relatively pristine state. One can observe them at play and performing whatever job they are assigned. This live version of the microbial ecosystem that is part of all humans will allow us to visualize and appreciate the invisible organ in the same way the neuronal networks in the brain or the calcium current that powers the heart have been recorded (Fig. 4.1).

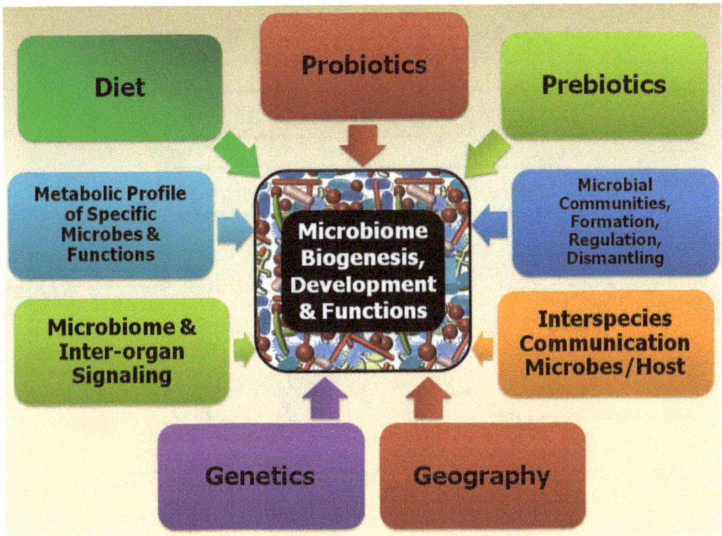

Fig. 4.1 Uncovering the biogenesis and regulation of the microbiome (the influence of various factors on the invisible organ, communication among the microbial communities and inter-organ signalling are uncovered).

These molecular microscopes will help us identify the precise functions this invisible organ is engaged in. Currently, the laudable task of having an inventory of all microbes residing within and on all of humankind and categorizing them according to various parameters such as age, ethnicity, eating habit, geographical location, and occupation are being undertaken. Then, it will be essential to understand the inner workings of the microbial communities that constitute the microbiome. Molecular fingerprinting and metabolite profiling coupled with data processing tools will no doubt unveil a gold-mine of information on the goings and comings of these invisible partners on which humans are dependent for their daily living. Just imagine in 10–30 years or beyond if discoveries continue at the pace they are currently being unravelled, the precise understanding of the biological contribution of the microbe to the body will be within reach. As the formidable exercise of fully classifying the microbial communities in various nooks and crannies in the body is executed, just imagine if the answers to the following questions are also available (Fig. 4.2):

1. How the microbial communities are formed, regulated and dismantled?
2. How are the community members chosen? What factors govern this process?
3. What chemical signals are utilized for inter-species communication i.e. between the microbes and the host?
4. How do they contribute to the development of human from birth and beyond?
5. Does the specificity of the microbial communities or the microbes that accompany us bestow on us character traits such as intelligence, longevity, athleticism, behaviour and physical appearance?

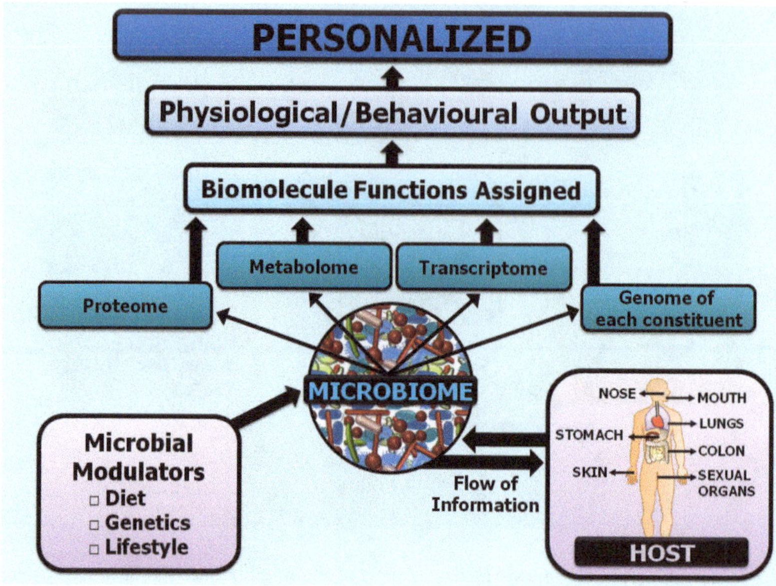

Fig. 4.2 The molecular elucidation of the microbiome (the role of each or set of biomolecules in individualized physiological function is known).

6. How external modulators like vaccines, diet, environment, probiotics, prebiotics and life-style can help mould our microbiome in a desired fashion?
7. What is the influence of epigenetics on the microbiome?

These are some of the questions if resolved will change our world faster than any discovery or invention has done up to this date. Unlike the machines and technologies we utilize to aid and ease our life, the molecular understanding of the majority of cells that constitute the body will have an unimaginable impact on our being and on what we can do with it. This is not like an external artificial prosthesis one summons in order to prop a failing organ like the hip or a stent in a blocked artery, but one can change everything one has come to know about the human body. Let explore how the world may/can be many years from 2018 when the cloak on the invisible organ is lifted.

4.2 Microbial Fingerprinting: A Potent Diagnostic Tool in Individualized Health Outcome

Examining the profile of the microbiome will become a standard diagnostic routine. It will be utilized to deliver health services in the same manner the blood group and blood contents are used today. This microbial fingerprint of each individual will be a repository of all the microbes residing on and within one's body

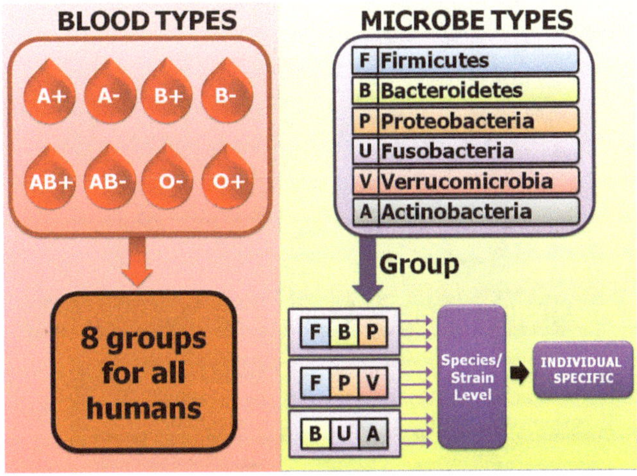

Fig. 4.3 A comparative view of microbiome and blood group classifications. (A,B,O blood groups and F,B,P,V,A,U microbial profiles)

just like the chromosomal or genetic information only few of the lucky ones can afford to be screened for at the moment. Health practitioners will be able to integrate the microbial ecosystem a patient possesses in his or her diagnosis and prescribe a therapy accordingly. Remember without the knowledge of blood types, the transfusion of this life-saving ingredient was risky only a few decades ago. The compatibility of donor and recipient is essential if this therapy is to succeed. Similarly, the nature of the microbial population will guide the decision on which microbial concoction will be needed to relieve a skin ailment or indigestion. It will also be possible to recommend microbial capsules, prebiotics or probiotics that will favour the promotion of the proper microbial ecosystem in diseased conditions where the microbiome has been perturbed (Fig. 4.3).

Currently based on the understanding of the microbiome, humans may be classified broadly into three main groups or enterotypes depending on their gut microbiota. Individuals with an abundance of *Bacteriodetes* may be grouped as Enterotype I while those having a prevalence of *Prevotella* are termed Enterotype II. Enterotype III is the classification reserved for people where *Ruminococcus* predominates. However, this biomarking technology is its infancy. Once, the specificity of each individual microbe residing in and on the body is sorted out at final microbial strain level, each human will have a unique profile as precise or even more accurate than the genetic information. External factors impact more rapidly the invisible organ than the traditional visible organs. These epigenetic changes can be easily picked up in the microbial communities the body is endowed with. The disruption of the microbiome is an important cause of diseases and the monitoring of the microbial population or its secretions can provide an early sign of any illness that may be brewing within. These tell-tale signs will be major diagnostic features of the future. These health monitoring processes will be less intrusive than the methods utilized today.

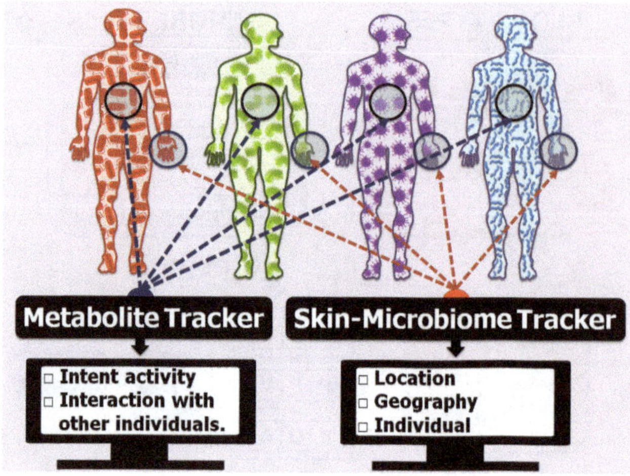

Fig. 4.4 Skin cartography: A reflection of the microbes and metabolites it harbours at a given time that are dependent on a variety of factors including activity and geography

Individuals can be tracked according to the microbial profile or the select metabolites they secrete. The skin cartography with the characteristics of the microbes and its secretions can be programmed in a detection device that can allow for the precise identification of individuals, and some diseases (Fig. 4.4). Furthermore, the behaviour or the intent of someone may also be extrapolated from these data. For instance, character traits can be individualized with the discovery of resident microbes and/or the metabolites responsible for a particular attribute. If someone is always in good-mood and jovial due to the propensity of his or her invisible partners to concoct the synthesis of say oxytocin, the good humour hormone, this feature can be tapped. On the other hand the ability of microbial ecosystem to interfere with serotonin formation that can result in depression may also be profiled. Just imagine these detectors sorting out individuals at large gatherings, at airports, during job interviews or when even selecting a leader. Currently, psychological tests are utilized to do such quality assurance with some degree of success as people can be programmed to lie or behave a certain way. Spies and terrorists are trained to evade these scrutinizing processes. However, the molecular microbial check will be relatively more robust and timely as it will reflect the biochemical markers the body elaborates before any biological response. A similar strategy can be employed to recruit the best athletes. If the microbiome one harbours can provide an edge in term of oxygen intake, this feature can be utilized to screen for the best performers. Some microbes residing in the gut can catalyze the synthesis of heme or the precursors of erythropotein, factors that stimulate more oxygen intake and can bestow on such individuals unique capabilities. Most of the athletes involved in doping scandals are usually trying to fortify their endurance by maintaining increased oxygen circulation in their blood system (Fig. 4.5).

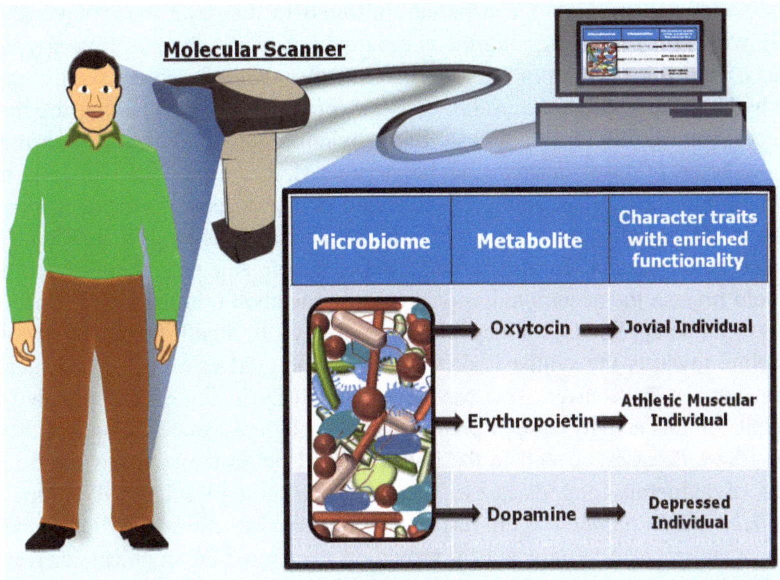

Fig. 4.5 Character modifying microbiome: (oxytocin, dopamine, erythropotein markers that may help screen individuals in hospitals, airports, and job interviews)

4.3 Obesity, Heart, and Gum Diseases

One common feature of these diseases is that they are all triggered by the disruption of the microbial communities that reside in the body. The uncontrolled production of toxins and the inability to guard against opportunistic bacteria are the primary factors that culminate into these ailments. Obesity is a major problem globally and has reached an epidemic proportion in many countries. It is known to be a drain on the health care system. Although a variety of factors such life-style, processed foods, and hormonal imbalance has been bandied around as possible causes of this disease, there has been also important data supporting the role of the microbiome in this ailment. Obese individuals have invariably a higher level of *Firmicutes* while people with less fat deposits tend to harbour an abundance of microbes belonging to the *Bacteriodetes* family. This shift in microbial ecosystem is characterized by a rise in toxins like lipopolysaccahride (LPS) that has been linked to weight gain and diminished sensitivity to insulin. The latter polices the level of glucose in blood. If the level of glucose is high, insulin marshals this monosaccharide to the tissues where they are trapped as glycogen and fatty acids to be utilized later. If the events that trigger the increase in *Firmicutes* or the specific strain of the microbe responsible for the production of the toxins can be manipulated to thwart this build-up, the health of obese individuals can be restored.

Prebiotics and probiotics can be administered to fine-tune the errant bacterial family which may be distressing the communal balance in the gut. Recently water extracts from the mushroom *Ganaderma lucidum* that are replete with complex carbohydrates have been successfully used as prebiotics in maintaining the microbial ecosystem in obese individuals. Trimethylamine oxide (TMAO) is an important predictor of how healthy the heart is. TMAO is produced by microbial intervention from choline, an ingredient found in egg or meat. Its elevated concentration in the blood is associated with cardiovascular diseases. Hence, identifying the microbe responsible for this product and or the enzyme promoting its synthesis will help reverse the accumulation of this toxic metabolite in the blood. The same strategy can be applied to counter fatty liver disease. In this instance, the diversion of choline towards the synthesis of trimethylamine (TMA) results in the increase of triglycerides in the liver. This can be controlled by inhibiting the microbes that participate in this activity. Use of prebiotics and probiotics can be the natural therapy against these diseases. A shift in the invisible residents in the mouth is the primary trigger of numerous oral diseases. The overpopulation of microbial groups like *Prevotella* and *Desulfobulbus* are responsible for gum infections and oral malodour. These organisms can be targeted by providing the proper environment for resident microbes in the oral community to thrive and eliminate conditions that are conducive to the intruding microbes. The inability to utilize oral metabolites and the intake of prebiotics that cannot be consumed by the invading species are two strategies to eradicate the microbial distress in the mouth. Of course, the discovery of the precise steps how the perturbation occurs is central to this natural remedy (Fig. 4.6).

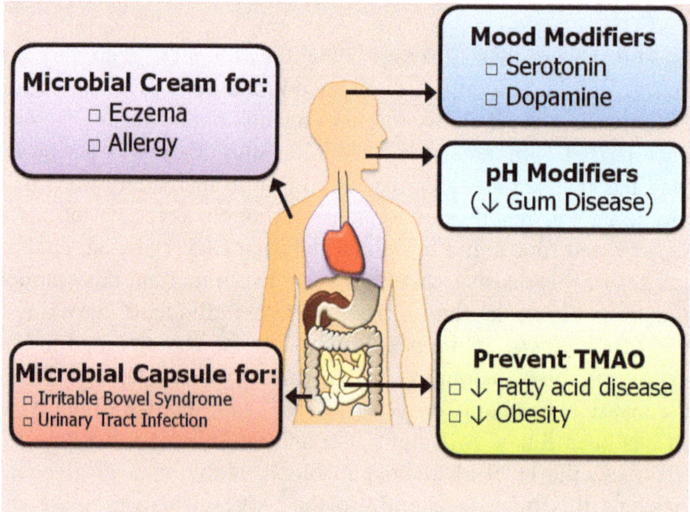

Fig. 4.6 Natural cure against diseases: a microbial remedy

4.4 The Skin: A Gold-Mine of Medical Record

The microbes residing on our skin can be a gold-mine of information as they can impart valuable medical data. They tend to be more sensitive to their surroundings and respond discernibly to the microclimates they are constantly exposed to. These microbes will be excellent clinical biomarkers for the prognosis of various diseases and also the environmental situation the host is confronted with. For instance, the nature of the microbial load at the site of a wound may specify when or if the wound can be healed or how the wound has occurred. If an object has been utilized to inflict a cut, the microbes may be reflective of this device or if the wound happened at a particular location like in a farm, this can also be deciphered. In this case, the nature of the farm can also be identified as the microbial profile of a horse, cattle or pig farm will be different. The metabolite cartography of the skin can be a good foreteller of one's health as these chemicals are produced by the resident microbes acting on the contents of the dermal landscape. A color-coded device can be routinely utilized to zero in on our health or the location we have been to. If for example diabetes is known to promote a specific microbe or a distinctive metabolite on the skin this can be the marker to be tracked. An analogous device is currently utilized to track dangerous chemicals we have on us when we go through security check at airports, seaports or stadium. Instead of these instruments being programmed for explosives or metals, they will be instructed to respond to select microbes and/or metabolites. This procedure will also be less invasive than the blood or urine sample that monitors one's medical status. Once this system is set-up, the microbes on the skin and their metabolites can become the gold-standard for screening diseases, identifying places we have been and may be pin-pointing the acts we intend to commit (Fig. 4.7).

Furthermore, creams supplemented with probiotics, prebiotics and select nutrients responsible to modify the dermal microbiota can be administered to promote working microbial communities or supplant the noxious ones. These remedies will be more beneficial to the host than the antibiotics and steroids that are routinely prescribed for eczma, psoriasis and other skin disorders. In the same manner as a stethoscope is employed to diagnose various body parts via the sound they emit, a device aimed at identifying the metabolites secreted by the skin will be a powerful tool to record the health status of an individual. Such a device will become an important addition at the disposal of health professionals as it will provide precise molecular information in a relatively non-invasive fashion. Envisage a situation when a teenager will quickly be informed of the cause of his or her acne and a remedy recommended even before the nasty facial spots flare up or a depressed individual will find out that the some of the neuro-active compounds are low due to the inability of the microbial dwellers to supply the precursors needed to produce these chemicals.

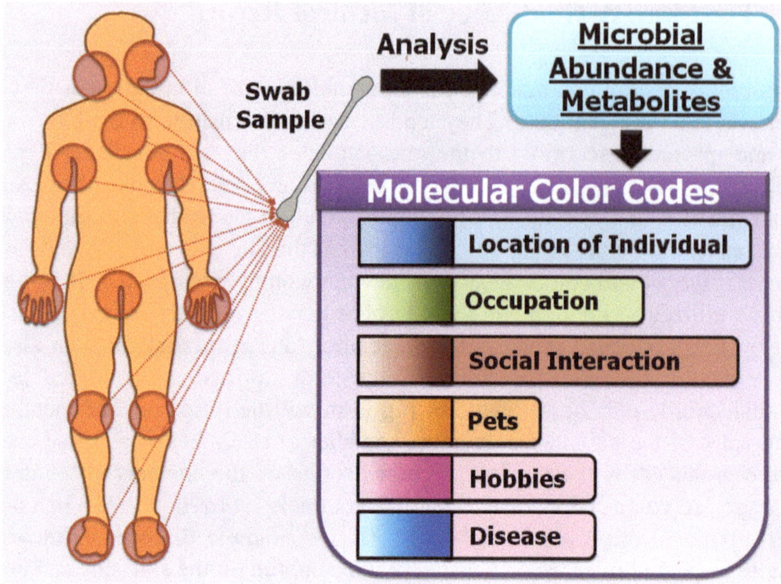

Fig. 4.7 The skin microbiome and metabolites: a gold-mine of medical data

4.5 Fortifying the Disease-fighting Ability and Drug Potency: The Pharmacy within the Body

The onset of numerous cancers is hastened by the inability of the body defence system to impede and arrest the growth of cancerous cells. These rapidly dividing cells then invade neighbouring tissues. The errant cells can be stopped on their track by activating the body's own chemical soldiers with the assistance of the invisible organ. Microbes like *Lactobacillus johnsonii* that nest in the intestine can be made to spring in action by cajoling the immune system to fight back and by neutralizing the tumor-inducing environment. Our microbial partners can also help by enhancing the activity of the medications we take. Their interactions with the drugs we deploy to combat diseases can render them more potent or ineffective. Their intervention is central to how the tissues or organs will see the drugs. The way the drugs are presented is decisive to how we respond to medications. Sorivudine, an antiviral drug is converted by gut microbiome into a toxin that can be fatal while digoxin, a drug aimed at treating cardiac complications is subject to microbial inactivation, thus decreasing its potency. The medication prontosil that is prescribed as an antibiotic is rendered effective due to its transformation by gut microbiota into an active component involved in arresting bacterial growth.

Melamine found in tainted milk in China led to the death of many infants. In an effort to inflate the protein content, milk formula powder was adulterated by this chemical. But unbeknown to these food fraudsters, the gut microbiome

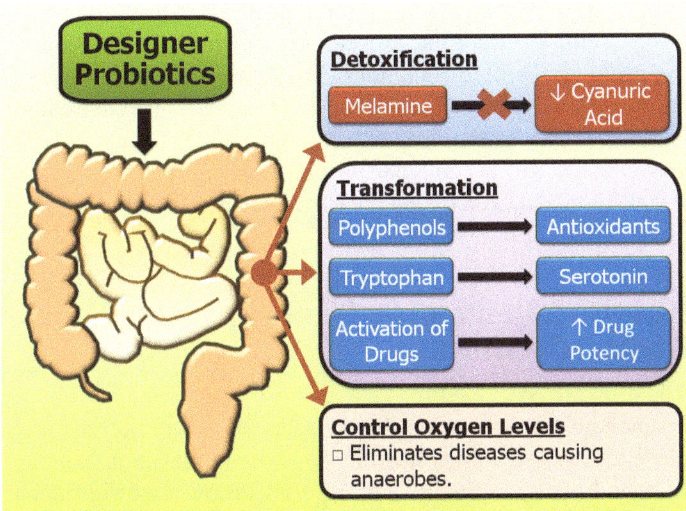

Fig. 4.8 The pharmacy within – potency of drugs modulated by the microbiome

converts melamine into cyanuric acid that then conspires to form kidney stones, a serious medical complication. Numerous thousands children were affected and even a few fatalities were recorded. The ability of the gut microbial ecosystem to transform drugs and also diet-derived bioactive products like polyphenols can have serious influence on our health. Hence, understanding the detailed features of the microbial community in the gut will help calibrate the administration of drugs, improve their efficacy and diminish their unintended impact. The current state of knowledge on drug interactions in the body only factors in the action of the liver. The role of the invisible organ though critical in this process is only now beginning to emerge. This knowledge will open the gate for more in vivo effective medications and eliminate the friendly fire the body is exposed to when we take drugs. Rarely, a medication acts precisely on the target and is devoid of unintended consequences on the adjoining biological systems. Our invisible partners will undoubtedly remedy these complications and bestow on drugs the real attributes they are designed for (Fig. 4.8).

4.6 Designer Babies and Biological Traits on Demand

The untangling of the molecular workings of the microbiome will open the prospect of acquiring desired biological attributes at will without the need of any genetic engineering. In this instance, the microbial ecosystem will do the job. The identification and elucidation of the parameters that lead to the establishment of microbial communities and their subsequent modulation by the host will lay the foundation to fundamentally change the perception we have of our well-being.

We will be able to nurture and nourish these controlling features in favour of the microbial ecosystem we want and impede any imbalance of the microbiota, an event associated with diseased conditions. The significance of life-style, prebiotics and probiotics in maintaining a perfect internal and external environment conducive to the proliferation of those microbes responsible for a better quality of life cannot be ignored; it will become part of our day-to-day vocabulary in the same way proper nutrition and exercise are the current mantra.

4.7 Character Adjustment on Demand: Microbial Contribution

Feeding, maintaining and renewing the invisible organ will become in vogue and will be equated to wellness. Microbial therapy will be crucial in helping our body function well and even in giving us an edge in life by bestowing upon us some unique attributes aimed at outperforming our peers. As the communication and the molecular language between the microbial communities and the host are unravelled, it will be within the realm of possibilities to tweak this partnership in such a manner as to promote the nurturing of uniquely desired traits in the host. These character adjustments can be initiated at the early stages of development. The placenta is not sterile but does possess microbial communities which interact with the embryo and the fetus. As the life-style and the eating habits of the mother are critical for the proper maturation of the baby so is the input from the microbial residents within. These formative moments are also assisted and interrogated by the presence of other microbial partners the mother has.

It is then quite possible that the development of the child to some extent is dependent on and responsive to the signals generated by the resident microbes. By learning about this very first encounter with the invisible partners, it may be possible to transmit desirable biological features onto a child; these attributes will impart a competitive advantage in the real world. For instance, increased neurogenesis may lead to a faster developing brain while increased production of vitamins will fuel enhanced metabolic activity. The ability of the microbiota in the placenta to prompt the synthesis of such blood components like heme may provide a child with a higher oxygen delivery system. This may result in heightened physical endurance, a well-sought trait in the sporting arena. The decoding of the chemical trafficking between the microbiome and the developing child may also help rectify some genetic errors or hereditary diseases that child may be predisposed to because of family genetic make-up. Instead of gene therapy or stem cell therapy, prebiotics or probiotics that confer unique microbial population in the placenta with the ability to eliminate the genetic malfunction can be administered. These in utero treatments will become a common practice (Fig. 4.9).

The future understanding of the microbiome will not only revolutionize how pregnancy can be managed but will also enable the promotion of physiological characteristics that can be tailor-made for each individual need. It is currently

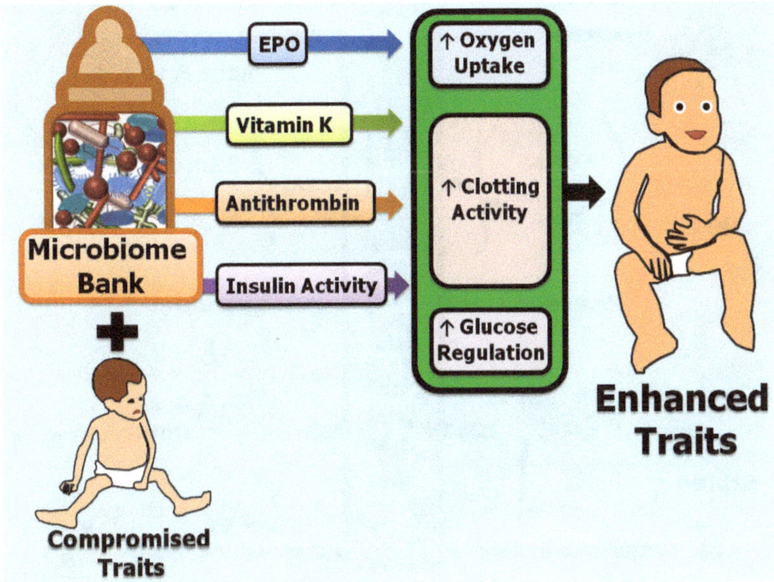

Fig. 4.9 Designer babies – microbes promoting unique traits

known that a shift in microbiota can trigger various diseases and rebiosis, the establishment of proper microbial communities can restore a healthy outcome just like in the case of irritable bowel syndrome. The introduction of *Lactobacillus* creates an environment that heralds the proliferation of good bacteria and the demise of the bad ones. Then, it stands to reason that microbial ecosystem can be harnessed to help marshal the synthesis of anti-thrombin and other disease-thwarting chemicals. Such a scenario will eliminate the risk of blood clot formation in an individual suffering from thrombosis. These clot-bursting microbial soldiers can either give instruction to synthesize the clot-disintegrators or produce the clot-gobblers themselves in manner analogous to how lactase is made available by microbes to individuals suffering from lactose intolerance. This improvement in the quality of life can be extended to a vast number of patients as this remedy can be applied to a variety of biochemical processes some individuals cannot execute properly (Fig. 4.10).

4.8 Microbial Rejuvenation of Old Age

Old age is a stage in human life that will benefit tremendously from the knowledge and technology associated with the microbiome. As we age our body is punctuated by a variety of physiological changes and abnormalities that start affecting the way our organs function. Can microbial partners be utilized to ease the transition from adulthood to life as a senior? Yes, it may be possible. There is a drastic reduction in

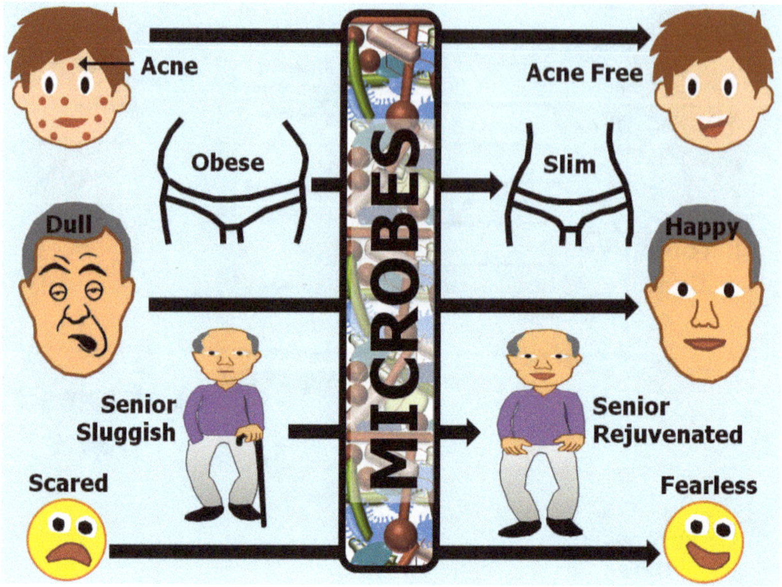

Fig. 4.10 Health promotion on the tap. A microbial boost for seniors and wellness in general

the abundance and diversity of microbial communities residing in and on the body as our age progresses. In fact, elderly individuals who are in poor health harbour least amount of microbes while on the other hand centenarians most of whom have a good quality life are flushed with diverse microbes as part of their invisible organ. Following new knowledge in this discipline coupled with the discoveries on the microbiome of centenarians, the microbial ecosystem in seniors can be adjusted so that they can participate in a fruitful life and not be condemned to sedentary living in retirement homes. By calibrating the microbial landscape to resemble that of an adult or even to someone who has lived beyond a century, it will be possible to make seniors more healthy. These microbial populations will promote the synthesis of various factors such as vitamins, enzymes, short-chain fatty acids and neuroactive amino acids such as tryptophan that are diminished as ageing occurs.

These new ingredients in the gut of seniors will coach the body physiology to respond accordingly and bring back biological functions that may have gone inactive due to the absence of these chemical cues to prompt them in the proper direction. The renewal of the microbiome can be fuelled by prebiotics or probiotics aimed at stimulating the growth of select microbial communities. Studies from disparate cultural groups where seniors fare very well biologically can provide clues to the specific microbes that need to be replenished. Similar global observations led to the discovery about dietary habits that were responsible for wellness right across national frontiers and cultural barriers. Including olive oil and fish that are rich in polyunsaturated fatty acids (PUFA) as part of regular meals has become synonymous to healthy heart. This observation was first reported in island nations or in countries where these two commodities were widely consumed. Hence,

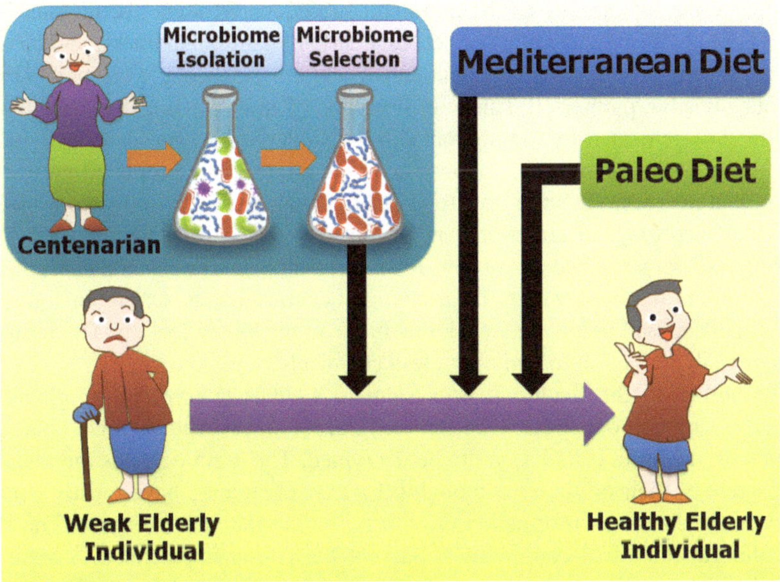

Fig. 4.11 Microbiome and rejuvenation of old age (mediterranean and Paleo diets promote healthy outcomes; Centenarian's microbiome rejuvenates the elderly)

finding a link between brain functions and microbiome will pave the way to transplant these microbes in individuals suffering from age-related dementia. In the same manner we now know that the Mediterranean diet is a tonic to the heart and the Paleo-diet may help with obesity, one will able to select the microbe or prebiotic or probiotic from a particular group of people somewhere in the world with the best health and longevity predictors. This will aid in rejuvenating the brain power or will have unique potency against any foible of old-age. Elders will be a group who will definitely benefit as the restoration of their dwindling microbial populations can be the magical elixir. Can the microbiome or the prebiotics from regions around the world with the highest concentration of seniors like in Okinawa provide a solution to ageing and enable life with dignity as one progresses beyond 70 years? Centenarians will also be an excellent candidate group in this quest to eradicate age-related physiological complications or will novel microbial communities or probiotics concocted in the laboratory come to the rescue? (Fig. 4.11).

4.9 How Microbes Can Control Behaviour: Mood and Brain Power – Microbes as the Brain Tonic

The significance of the gut-brain axis is only now being recognized as a pivotal component of the signaling system in the body. The gut is the home of most microbes in the human body and they produce a wide variety of chemical stimuli that dictate the functioning of the brain and subsequently human behaviour. They play a central role in controlling appetite, satiety, pleasure, happiness, and anxiety. Whenever you have craving for chocolate or any other delicacy, you may blame it on your invisible partners!!! This sub discipline of how gut microbes or how the volatile cues generated by the microbial skin dwellers affect our daily activities is still at very rudimentary stages. However, if the microbiome-propelled network in the gut that instructs the brain how to respond is unearthed, the ability to transform behaviour, neurological afflictions and society at large will be within reach. Yes, currently we rely exclusively on the innate neurotransmitters and their influence on the brain to predict a variety of behavioural characteristics. Oxytocin and dopamine generated within are the switch that turns on the happy feeling. The influence of the invisible organ in modulating these chemicals is not known.

The ability of the microbial ecosystem to synthesize tryptophan, glutamate, dopamine, and neuropeptides has been recognized, however their direct communication with the brain cells has yet to be decrypted. The brain decodes these signals irrespective of their origin even though till today researchers have mostly concentrated on the chemical information originating within the visible organs. The input from the gut microbial communities has not been fully appreciated, a feature if fully elucidated will completely transform people's lives. Modulating the levels of these neuronal commands can endow the brain and the body desired physiological properties and unique attributes. Can it be that as we age the reduction in microbial abundance and diversity in the gut is responsible for all the mental complications we suffer from? Can the microbial ecosystem be adjusted to promote the synthesis of these signals that will then have a healthy outcome for our seniors? It is entirely within the realm of possibilities that we can charge the brain and fine-tune its behavioural intervention to suit any desired physiological response. The microbial profiles of individuals and or groups of people residing around the globe with unique mental attributes or behavioural comportments can be prized ecosystem that can be exploited.

Individuals with exceptional intelligence quotient (IQ), or those having discovered amazing technologies or nobel laureates will be excellent starting points to unravel the microbiome connection to intellectual aptitude. The identification of links between microbial communities and mental faculties will lay the foundation for microbial transplants responsible for human behaviour. Is there a link between sleep pattern, microbiome and happiness? For instance if obesity is attributed to a microbial ecosystem that regulates leptin, the appetite hormone, individuals who want to lose weight can be given prebiotics or probiotics favouring this community (a recent study has revealed that countries like Timor has the most amount of people with the lowest body mass index (BMI). This can be a fertile ground to initiate such a search project); oxytocin forming microbiota can be utilized to bring happiness to individuals who are predisposed to be sad. Anger, irascible behaviour, mood-swings, and killer instinct can all be inoculated through this microbial therapy. One can even program individuals to do insurmountable feats. May be looking at the microbiome of dare-devils, stuntmen and ropewalkers can revealed the role microbes have in promoting these traits. This will be a gold-mine

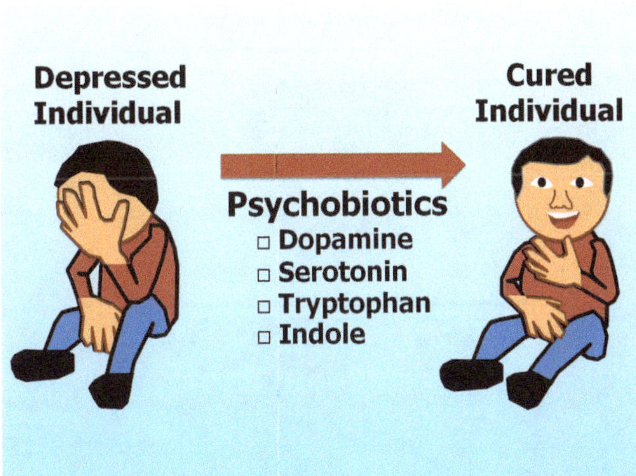

Fig. 4.12 Microbiome and behaviour modifiers

for the microbiome hunters and a delight for the common person who after receiving a dose of these microbes can partake in these amazing feats. Here there is no drug but only a shot of the right assemblage of microbes. It will be a perfect birthday gift for those of us who want to indulge in some unique daring activity on a special day (Fig. 4.12).

4.10 A Microbiome Bank for the Health Care Enterprise

As of now the influence of the microbiome on health outcomes is not factored in and all tests and therapies administered to remedy diseases are dedicated towards the visible organs. Even the medications one takes to rectify an ailment are given without factoring the possible intervention of microbes residing in the body. Only when drugs become toxic following microbial input, then the alarm bell is raised. Oh yes, these invisible partners can sabotage the properties of medication by either enhancing its therapeutic potential or neutralizing its medical impact or augmenting its toxic influence. However, all this will change when the genesis, evolution and functions of this invisible organ are deciphered. The harnessing of the assets this organ is laden with in regard to body's well-being will completely change health care education and delivery. The intricacies of the microbiome will become part of the normal health curriculum in the same way the fighting of disease-causing bacteria is currently taught.

Remember the blood groups that are an essential component of medicine today was not categorized until the early twentieth century and the Rhesus factor, the + and – signs attached to the blood imprint (ABO) was only revealed to the world thirty years later. The microbes within and on us will be classified according to the functions they perform, the community they live in, the part of the body they

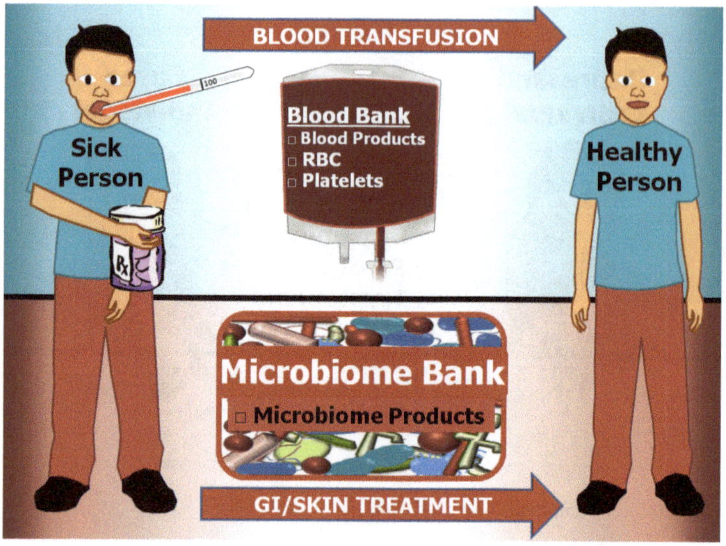

Fig. 4.13 A microbiome bank: a health treasure-trove waiting to be utilized

reside in, the essential chemicals or signals they generate or to the unique attributes they impart in a manner analogous to stem cells or to various other yet to be construed parameters. Healthy individuals will store their microbial ecosystem in microbiome bank the same way blood bank stockpile blood and its various components (Fig. 4.13).

For instance, following a surgery, antibiotics may be administered to promote the recovery of the body. The natural microbiome of the patient may be compromised after a medical intervention; in such a situation the microbial ecosystem stored in the microbiome bank can be prescribed in order to assist the body to get back to normalcy faster and thwart any new infection, a common occurrence in recuperating patients. Prebiotics and probiotics promoting the patient original microbiome may also be helpful. This easy to implement procedures can be real life-savers especially to vulnerable groups like children, elderly and people with specific physiological needs. Patients whose health has been compromised due to the imbalance in the microbial population, can be given microbial transplant from the microbiome bank. This can apply to ailments like irritable bowel syndrome and fatty liver disease. Babies who are born premature or delivered by C-section tend to inherit only limited amounts of microbiota from their mothers, a situation that make them more susceptible to asthma, allergy and other abnormalities. To correct this deficiency, the mother's microbial ecosystem can be swabbed on these babies or microbes from the microbiome bank with unique physiological properties belonging not to their biological mothers or microbiome from matching donors can be also administered. In some instances, feeding them the prebiotics that promote the proliferation of the desired microbial landscape can also become an effective strategy.

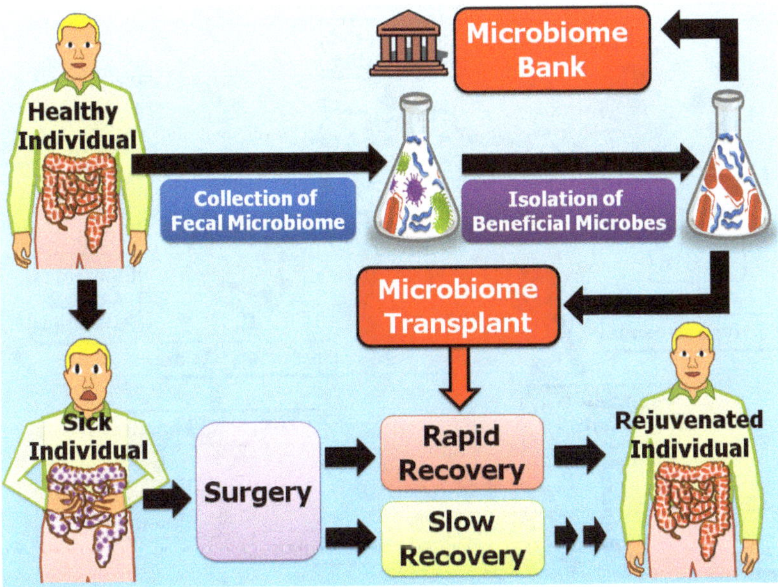

Fig. 4.14 Fecal transplant – an evolving therapy

Fecal transplant is one form of microbiome therapy that is being actively pursued. This treatment is routinely being given to patients who are infected with drug resistant *Clostridium difficile*, that is stubbornly tenacious and hard to eliminate with antibiotics. The administration of microbial communities from the stool of healthy individuals to combat obstinate infections is gaining lot of praise and has become the go to remedy in many hospitals even though it is only at an experimental stage. As this therapy is optimized, it will become a standard procedure to combat contagious microbes especially in elderly population whose immune system tends to be less responsive to infections (Fig. 4.14).

In Asian veterinary medical practice, utilization of microbes to fight diseases in cattle has been successfully applied for thousands of years. The introduction of rumen fluid from healthy cows has been an excellent remedy for cows that are unable to ruminate. Young calves inoculated with maternal cud in order to stimulate their digestive system or hatched chicks smeared with chicken feces in order to increase their survival by prompting the immune system are other illustrations of the success of microbial therapy. Although fecal transplants are gaining traction, the ability to readily monitor the microbial ecosystem and their components in the human stool will greatly speed the adoption of this therapy. The microbiome can also be mined for a single micro-organism or a consortium of microbes dedicated to a desired physiological task. This will lead to the creation of unique microbial systems that can be reproducibly concocted and will be devoid of undesirable pathogens or viruses. This will give rise to microbiological industrial ventures dedicated to supplying these medical products. Such a biotechnological

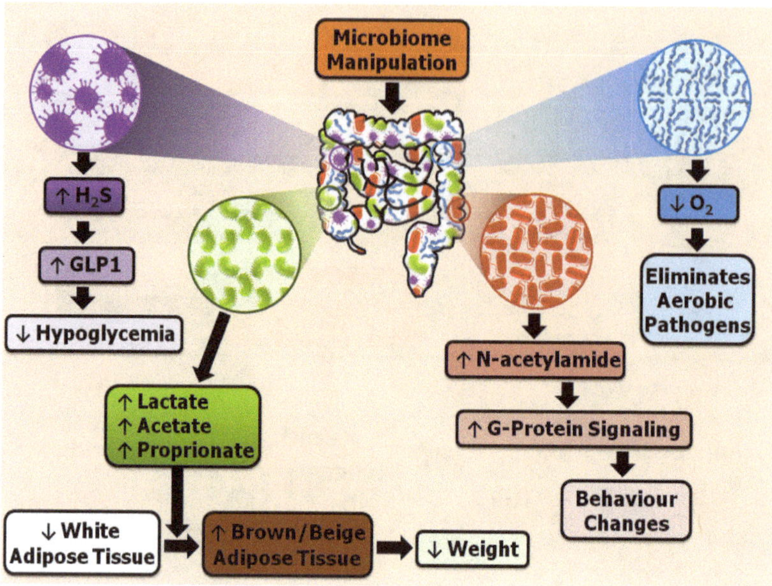

Fig. 4.15 Microbiome manipulation and healthy living

innovation will quickly earn public support and its products can be dispensed in the same manner analogous to stem cells, organs or tissues. It will rival these established organ banks and its constituents will be easy to administer. Microbiome engineering will herald a new approach to health care and will lead to a dramatic shift in patient well-being. It will provide an additional tool just like blood components in helping improve the quality of life (Fig. 4.15).

Just imagine where health care sector will be without the blood bank. Millions of lives are saved yearly through blood transfusion and the recovery of the loved ones is rarely compromised because of lack of any blood-related therapeutics. Even though the first blood transfusion was performed only in the early twentieth century; the first blood bank usually attached to hospital did not happen until 1936. Shortly after national banks were established and blood collection drive was instituted. Till today, new technologies in the isolation and separation of this vital fluid are being introduced and new formulations of biological products are being uncovered. Organ banks have also become an integral fabric of the health delivery system. Solicitation of organ donors is an important societal endeavor and people are encouraged to sign their donor card, a wish that is acted upon following their death.

A similar medical necessity will emerge with the microbiome bank and this will become an important hub for treatments and ingredients derived from human microbial communities. For instance, a microbial consortium dedicated to the synthesis of vitamin K may become an excellent remedy for individuals lacking this blood coagulating factor; a person suffering from depression can be given an

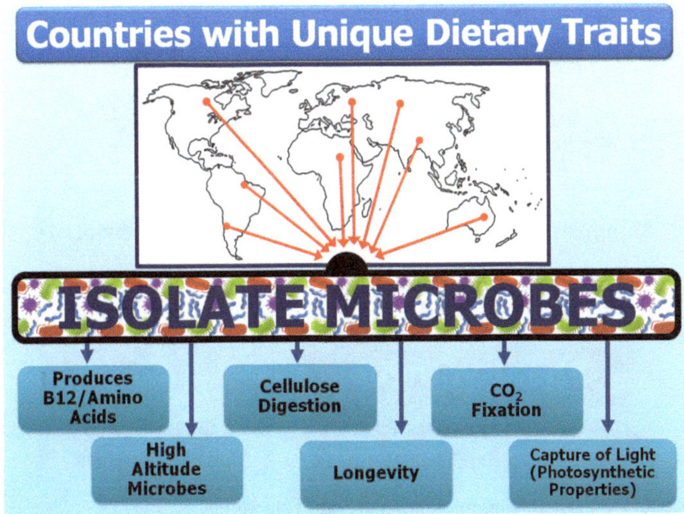

Fig. 4.16 Global search for biological function-specific microbiome. (microbial solution to health issues)

inoculum with a dopamine-generating characteristic. A bank with human microbial population and compatible microbial ecosystem will become a normal fixture in a hospital or clinical setting. People who can afford it may even opt for private microbial banks where their beloved microbes can be safely nestled for future use or for future generations if they are deemed to harbor some unique properties. Where were sperm, human egg or gene banks just a few decades ago? Well almost everyone is aware of these institutions and some entrepreneurs are making a fortune out of these. A similar motivation will fuel the emergence of commercial microbiome-based entities (Fig. 4.16).

4.11 Microbiome: The Health Revolution in Waiting

Once, all the intricacies of the invisible partners have been decoded, the microbiome will be a gold-mine of biological assets that will be minted into a healthier world. With the discovery of unique microbiomes across cultural and geographical boundaries, and the unraveling of either a microbial consortium or a microorganism dedicated to a well-sought after physiological function, the contribution of the invisible organ to the well-being of human and the health system will be tremendous. As an example, just consider what the identification of a gut-residing microbial community that can degrade cellulose can do to the world-food order. Almost like with a magic wand, the clouds of hunger and malnutrition that hang over a good portion of the world will disappear; there will no longer be children begging for foods and food relief agencies will have to be engaged in other

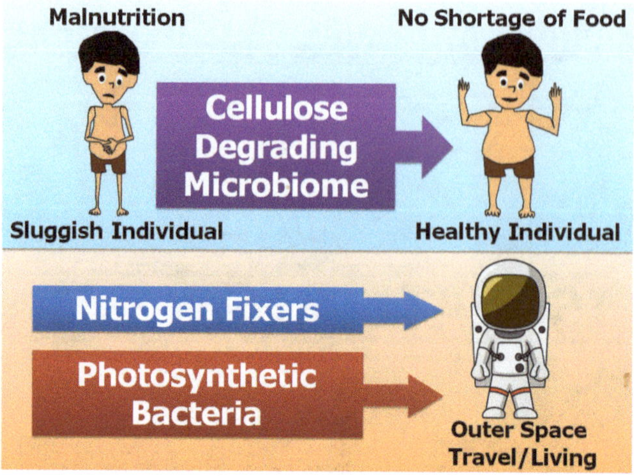

Fig. 4.17 Cellulose metabolizing microbes and end of starvation

humanitarian activities. Treating people or helping individuals acquire unique biological traits with the assistance of specific microbial community or select individual microbe will be a common practice in patient care and in promoting a healthy living. It took numerous decades to master blood transfusion and to utilize its various components to cure diseases. And today we cannot imagine a hospital without the therapeutic power blood provides for the success of the organization. The microbiome will reach a similar status in the healthcare industry in not too distant a future (Fig. 4.17).

Although the microbiome science is currently at what one may term the 'Van Leeuvenhoek era' that is an epoch dedicated to identifying and cataloguing the constituents the microbiome, we are just at the doorsteps of major findings that will revolutionize the understanding of how the body ticks with all its organs. Following the initial observational stage, the delineation of how these invisible partners operate will pave the way for numerous applications. The evolution of the microbiome with the physiological and biological development will guide and inform the establishment of new remedies that will be critical in all aspects of human life. At the present time the pharmaceutical industry is still based on the structural manipulation of chemicals in an effort to find cures for different ailments. However, this approach is marred with numerous side-effects and often does not effectively accomplish the desired goals. The reduction in the potency of the drugs and their side effects are two major drawbacks. Understanding the fine workings of the microbial ecosystem will enable the inclusion of the invisible partners in shepherding the drugs in the proper direction to optimize their potency. Hence drugs can be made more powerful, less toxic and target-specific with the intervention of the microbes. Furthermore, the manipulation of the microbial communities themselves or their products will enable the improvement of desired physiological functions. And as these are derived from the body's cellular

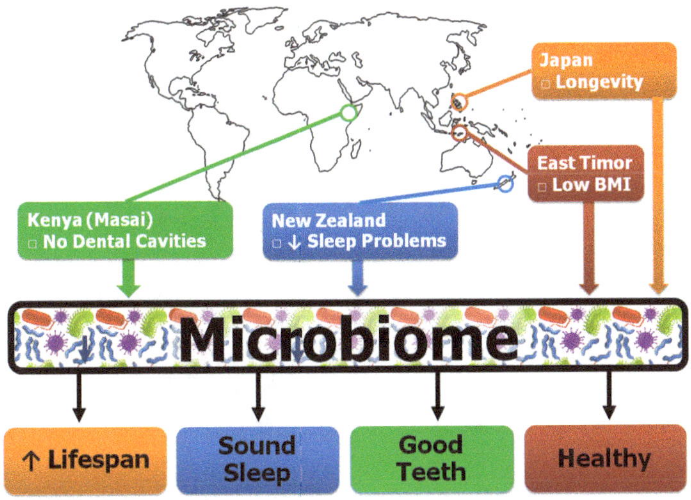

Fig. 4.18 Global microbiome screening: search for unique physiological property

components, side-effects will be relatively negligible. It will be like treating diseases with blood products but in this instance the establishment of a particular microbial ecosystem will have more lasting effect and can be easily maintained through intake of prebiotics or probiotics (Fig. 4.18).

Let's consider the case of vegetarian diets. These are perfectly nutritious and healthy meals that need to be well-balanced, as essential ingredients like vitamin B12, some minerals and amino acids may be lacking. To remedy this situation, the gut microbiome can be tailored to supply these essential nutrients. This type of designer digestive tract can be programmed for individuals with various kinds of dietary habits. People who have a soft spot for fatty foods or a rare beef steak can take a microbial ecosystem designer pill to train their invisible partners in the gut so that they can indulge in their eating pleasures. Digestive tablets with enzymes like trypsin and lactase are prescribed to individuals who suffer from digestive problems. In the elderly segment of the population where this problem can be acute due to the inability of the organs to synthesize these enzymes in sufficient quantities, microbial therapy will be an added benefit as it will obviate the need for daily pills. The ability of the microbiota to change physiological functions can be a boon to athletes. Instead of utilizing steroids and other drugs to boost their endurance and their performance, microbial re-engineering can become handy. It is well documented that an increase in the hormone erythropietin (EPO), keeps the body flush with oxygen especially when the need of this commodity is critical during extensive physical activity. This can boost one's competiveness and provide a clear path to the medal podium. Reconfiguration of the microbial population with the ability to directly or indirectly regulate this hormone can be the difference between a bronze or gold medal in the Olympics games. The caveat here is that the rules of the competition may evolve with this microbiome technology (Fig. 4.19).

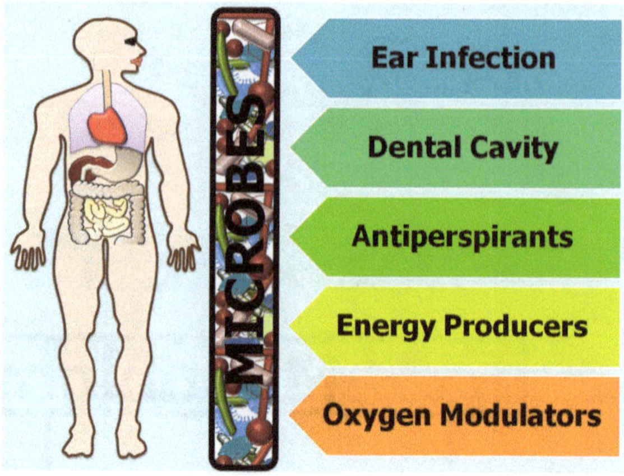

Fig. 4.19 Microbial solutions to daily living (these tailor-made microbial supplements can help solve these problems)

4.12 Xenomicrobiomes and Human Wellness

Although human microbiota and its products will be the primary contributors to the microbiome bank, it is very likely that microbial communities from other species will be a valuable addition. Blood banks stockpiling animal blood and its derivatives are well-established. These institutions supply products that fulfill the veterinary market but also contribute to demand geared towards human health. Xenotransfusion i.e. transfusion of animal blood into human though uncommon feels the void when there is a scarcity of blood during natural calamities when shortage of human blood is acute. In fact red blood cells from pigs have lot of similarities to humans. Animal blood products are routinely utilized for diagnostic and scientific experiments. Once, the microbiomes of pertinent organisms are catalogued and the biological functions identified, their components can be packaged to ease human sufferings and for the betterment of society as a whole. Currently, the practice of pet therapy is in vogue and has gained wide acceptance in many institutions. In retirement homes, hospitals and universities, this non-invasive remedy has been and is being invoked to relieve the daily stress some individuals face. While in hospitals the convalescence period has been shortened, seniors in retirement homes have been prescribed less medications needed to go through their day to day routines. In fact, seniors are known to look forward to their daily interactions with the pets. Examination blues and living away from their homes, the two hurdles university students are confronted with have been mitigated by the involvement of pets in the mix. The magic behind this pet therapy is the perception that pets provide a psychological tonic to individuals patting them or holding them. It is quite reasonable to assume that this bond is triggered by the exchange of microbes between the protagonists.

In fact societies where animals in close contact with humans, there is a horizontal transmission of microbes i.e. human microbes are observed in pets and vice-versa. Mice living among humans have their immune systems resembling humans compared to laboratory mice supplied by breeding companies. This may be the secret contributing to the jovial environment induced during such an interaction with pets. Even compatibility among family members living together may have a microbial twist to it. Remember the Stockholm and Lima Syndromes; Stockholm syndrome is a term used for the bond that develops between captors and hostages. In this instance, the hostages start developing a likeness or sympathy for captors. During Lima syndrome the opposite phenomenon is observed i.e. the affinity and the bond of kidnappers for the captives increase and they start agreeing on various issues. Although different psychological postulates have been bandied around to explain these behavioral transformations, the influence of microbiota residing in and on the participants has never been taken into consideration. May be the knowledge on microbiome was not mature enough when these phenomena were first observed. Now with the innovative technologies capable of visualizing the invisible world that is part of us, it will be relatively easy to decipher how the horizontal exchange of microbiome communities may be contributing to these unexpected changes in attitudes. Could the microbial ecosystem and their components be at play? If there is a link, this will help law enforcement officers to crack cases or even help terrorists reveal their hideous machinations.

The exchange of microbial partners between animal lovers and their pets may play a role in the affection to each other. Hence, it is not surprising that Eden House, an establishment based in the USA catering to seniors is promoting the presence of pets in these residences coupled with the regular visits from schoolchildren and kids. The objective here is to make this living space as natural as possible i.e mimicking the environment one comes across daily. Again, the mixing of microbial ecosystems may be at work. The people frequenting or residing in Eden House have a better quality of life and live longer than patrons living in a run of the mill retirement home. Once the molecular knowledge on these microbe-based occurrences is fine-tuned, the microbial communities and their components can be banked and utilized for betterment of society. Microbes may even have a role in therapeutic gardens, rehabilitation greenhouses and horticulture therapy. These can then be given as supplements to individuals who are confined to a limited space due to various circumstances. Just imagine how an individual trapped in his/her bed will feel if he or she can experience the joy of gardening with the help of these garden microbes delivered in a spray bottles (Fig. 4.20).

4.13 Agro-Food Business: Foods Dedicated to the Microbiome

Agro-food business will evolve in order to adapt to the new information pertaining to the microbiome. Agricultural practices will need to response in such a way so as to produce foods that will meet the expectation of the invisible organ. Consumers will demand products that will enable them to modulate their microbial communities responsible for the most desired properties and to thwart the proliferation of microbes dedicated to wreak havoc on this fine microbial balance.

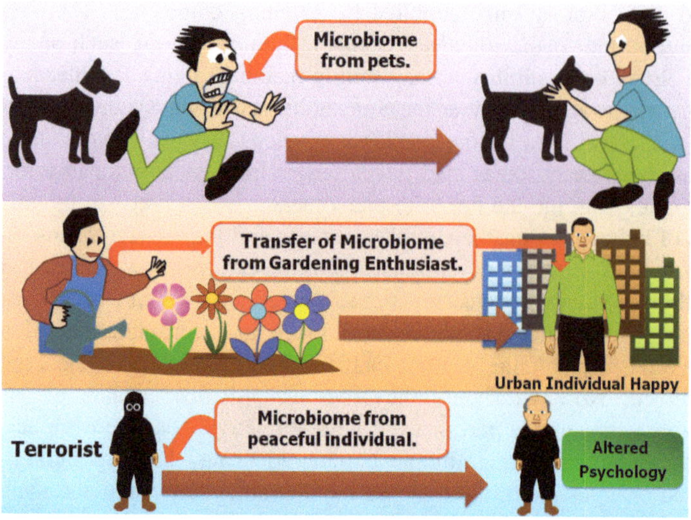

Fig. 4.20 The xenomicrobiomes and societal transformations (microbial transfer can help urban dwellers, pet phobia and terrorists)

Agricultural goods and processed foods will not be only a source of nutrients but will also be fertilizers aimed at the microbiome. These dual requirements on food products will force the food industry to come up with innovative solutions. Even now prebiotics and probiotics are being included as integral components of the foods we consume. Yogurt with enriched life microbial cultures and bran encrusted bread are a common staple on the grocery shelves. There will be a dramatic increase in these products and more importantly these nourishments tailored to have a direct impact on the invisible organ will be function-specific and time sensitive. Prebiotics designed to promote a select microbial community known to trigger the production of say alcohol dehydrogenase, will enable alcohol-intolerant individuals to enjoy at a party without the embarrassment of having an adverse reaction to alcoholic drinks. Foods aimed at increasing the microbial population responsible for the synthesis of EPO, a hormone that ensures the tissues are flushed with oxygen will be a boost to those undertaking strenuous exercise, in a manner analogous to marathon runners who partake in carbohydrate-rich meal (pasta/rice) before a race so that they do not run out of readily accessible glucose. The similar nutritional uplift can be provided to microbes promoting the synthesis of the brain tonic, lipoic acid. This will be music to the students cramming for their final examinations. Creating delicious edibles with uniquely crafted physiological properties via the intervention of the invisible organ will become a cause celebre of agri-food business in the same manner organic foods are to environmentally conscious customers.

The microbiome will become the darling of this industry. It will not resort to marketing sugar-laden snacks aimed at the taste buds or other supplements earmarked at our visible organs with complete disregard to our invisible partners. This strategy to target the microbiome will also curb obesity, metabolic diseases and mitigate malnutrition. Remember the banishment of trans-fats from foods and restaurants that is being instituted in all corners of the globe. This toxin found in foods is a bane of the heart. Any edibles inflicting damage on the microbiome will be readily relegated to the waste baskets. In regions of the world where the population suffer from lack of vitamins and minerals, foods enriched with prebiotics and probiotics with the instruction to commandeer either the synthesis or absorption of these essentials, will be a welcome remedy. Iron-deficiency in some part of the world is tackled by the consumption of brown salt i.e. salts enriched with iron minerals. The ability of some microbial community to secrete simple sugars from complex plant-based carbohydrates can help alleviate the global hunger problem. Physiological manipulation on demand will be a reality without the input of genetic engineering or stem cell supplementation but by eating tailored-made food. These designer foods will only spur specific microbial ecosystem to perform the coveted task.

4.14 Designer Prebiotics: Foods with Unique Functions

Furthermore, disparate prebiotic properties can be incorporated into a single food or fruit. For example, the virtues of an apple and a banana can be encapsulated in one fruit to the delight of the microbiome. Even the active ingredients like the FOS, GOS and XOS can be provided as pills. Remember the occurrence of a unique sialylated component in human milk has been shown to be the deciding factor between healthy and malnourished infants. At the present time, most of the animal or fish proteins we consumed are obtained from industrial farming. To fatten chicken or hasten the growth of salmon, antibiotics and steroids are often utilized. This mode of rearing protein-rich foods the world yearn for carries lots of risk as these chemicals can end up in the body. Precocious menstruation and the widespread presence of antibiotic resistant bacteria are among some of the problems that can be traced to this practice. This situation is being further compounded by the rise of the emergent economies like China and India where demand for protein-laden foods are skyrocketing. There is a solution; if the microbial ecosystems of these organisms are uncovered, feeding them prebiotics and probiotics will achieve the goal of plentiful protein-rich food but without the headache of the antibiotics or steroids. This will be an innovative norm in the agri-food business as the health and financial benefits will be too tantalizing (Fig. 4.21).

4.15 Human Microbiome Prospecting: A Walking and Talking Gold-Mine

Mining the invisible organ will be another endeavor that microbiome technology will inject in the economic discourse. A complete full-fledged commercial enterprise will be based on this and various nations will be competing ferociously to

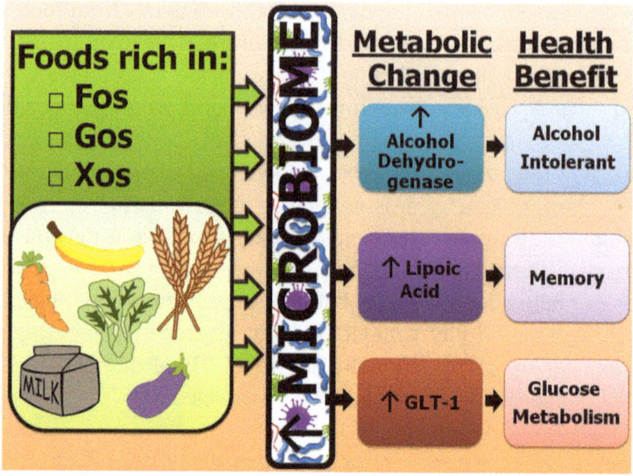

Fig. 4.21 Designer foods (prebiotics and probiotics with unique physiological attributes)

master and protect these wealth-generating ventures. The human body and most other living organisms are teeming with microbial communities whose functions are yearning to be identified, deciphered and then tapped for societal good. When one looks at oneself in the mirror, just think of the invisible microbial dust that blankets one's body inside out. These tiny workers are performing amazing task either as a community or even alone, these activities can be tailored for the betterment of human life. For instance, almost everyone has been exposed to the malodor emanating from co-travellers in a bus, metro, train or even co-workers in the office? Well, the remedy to this discomfort is within reach and it is not antiperspirant or odor camouflaging substance. Recently the culprit behind this foul-smell has been apprehended and the microbes belonging to the group *Corynebacterium* spp are the main perpetrators of the noxious scent. They are more abundant in the armpit of smelly individuals while the non-smelly underarms are populated by *Propionibacterium* spp.

The thioalcohol molecules that are responsible for this phenomenon can be debilitating. Individuals harboring these bacteria tend to repetitively utilize antiperspirants that can render the situation worst as they become fodder for the offending microbes. Transplanting the non-odor producing microbes can quell the stench or nanocapsules infused with these microbes can be helpful or even wearing cloth sprayed with these microbes can impede this smelly ordeal. It may be uncommon but our own microbes can rescue us from infections by invading microbes. One such example is the use of wax in the ear. Usually one ear is infected while the other one remains an infection free zone. The swabbing of the earwax from the healthy ear into the one with the infection can bring relief and cure to the patient. These two examples just reveal that we are just scraping the tip of the iceberg as far the applications of the treasure hidden within us and very likely in other organisms are concerned. This treasure trove unseen by the naked eyes will be unlocked and countries that will get access to these miniscule miracle workers will clearly have an edge and most probably transform society the way we know it. The microbiome is constituted of nanofactories engaged in generating unique goods that are waiting to be tapped and tamed.

4.16 Nanofactories for Desired Biological Functions

Microbiome prospectors will have a fertile microbial laden landscape to explore. The human ecosystem will be an excellent territory to map and lay claim on its most lucrative invisible habitants. These microbial partners can change our physiological functions, extend our life, make us athletic, eliminate our bulging bellies, enable us to diversify our nutritional habits and extract the maximum benefits from our food intake. Scouting the microbial ecosystem in societies and individuals living in unique geographical locations or having very peculiar dietary habits or engaged in distinctive life style will divulge stunning findings on microbes capable of biological wonders. Exploring the people in the Amazon or in the African tribal regions or the Aborigines in Australia who rely heavily on plant-based diets to fulfill their energy needs will be a fascinating scientific adventure. It is quite possible that their guts have adapted to this nutritional environment and have evolved with the invisible partners seeking refuge in this organ. These microbes may have the ability to supply the host with amino-acids and other essentials in order to maintain a healthy life-style over centuries.

The microbial communities can be mined and administered to individuals who suffer from alimentary deficiencies and can also be engineered to convert complex carbohydrates into amino acids. These microbial colonies must also be harboring nitrogen-fixers and will be an important contributor to the protein need of humans. Individuals who are part of the Masai tribe for instance, have almost no dental cavities. Can this be attributed to the oral microbiome? The individuals prone to cavities will benefit from this. Identification of people with voracious eating habits without increasing their BMI will become a prime prospecting biological landscape for microbial communities aiding in this task. The obesity epidemic will be alleviated. The global food crisis can be remedied if microbial ecosystems with attributes to convert cellulose into glucose or to harness the sunlight are identified. The latter will be huge as it will fulfill human desire to explore the outer galaxy. These microbes can be sprayed on the skin or established in the gut without the need to stockpile foods in the spacecraft. Microbial partners participating in such a task have been found in the organisms living in hydrothermal vents. Any nation that is successful in obtaining this biotechnological proprietary right will undoubtedly alter the world power structure. People living in elevated altitudes can provide us with a wealth of microbial assistants mitigating the dangers of life in the absence of optimal level of oxygen. Dwellers of arid lands will most likely have a cache of bacteria dedicated to supporting a life style where there is a dearth of water.

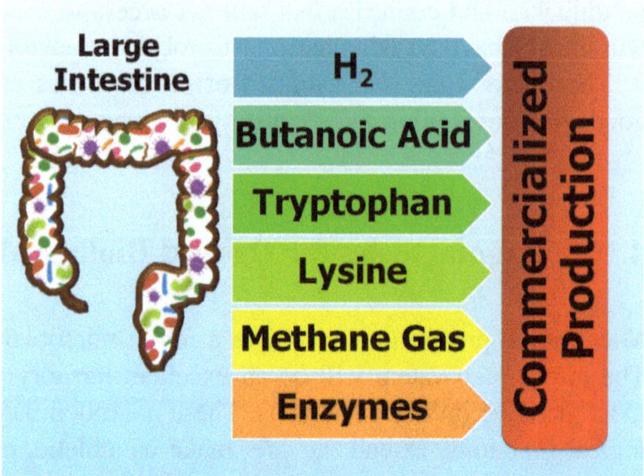

Fig. 4.22 Microbial prospecting: endless possibilities (human intestine is rife with microbes having commercial value)

Microbiomes from other organisms will also possess a bountiful of microbial life involved in intriguing physiological processes or generating valuable products. In this instance animals that are closely related to humans biologically will be obvious candidates. Microbiome of gorillas and other primates may harbor select microbes that can readily be adopted by humans. However, the life-style of organisms in extreme environments or feasting on hard to digest foods can be fertile ground for microbiome prospecting. As a starter, the wood-pecker, the termite and the cow can be a gold-mine of microbial opportunities. These organisms readily metabolize cellulose, lignin and complex carbohydrates derived from plants. These microbial communities will be prime target for commercialization as they can transform the normally indigestible carbon compounds into simple and easy to utilize nutrients. The unearthing of a microbial community programmed to metabolize cellulose will be a major boon to world food policy framers. Just like how gene and genetically modified organisms are patented, the discovery of microbial systems with all these marketable attributes will be treasured for their monetary values. Some tribes living in a far distant land can become a gold-mine of much sought after microbes in same manner geologists survey the globe for mineral deposits and gene hunters seek rare genes to commercialize (Fig. 4.22).

Microbes residing in the gut can be engineered into nanofactories designed to produce energy, enzymes and drugs. Bacterial communities residing in the colon are known to produce such energy-rich chemical as propionic acid, butanoic acid, butanol, and H_2. These microbial machines can be harnessed for industrial production. How a unique microbial community discovered in humans or for that matter in any other organism with ability to produce H_2 can transform the energy landscape of the world. Planet earth will have the cleanest energy around and the beauty of this process is that only the highly valued water is the by-product. Cars will inadvertently become water-generating machines. The same innovative and revolutionary ideas can be applied to the production of enzymes that can degrade complex carbohydrates into simple sugars or hormones that can fulfill various societal needs. The economic value from exploiting the invisible organ most

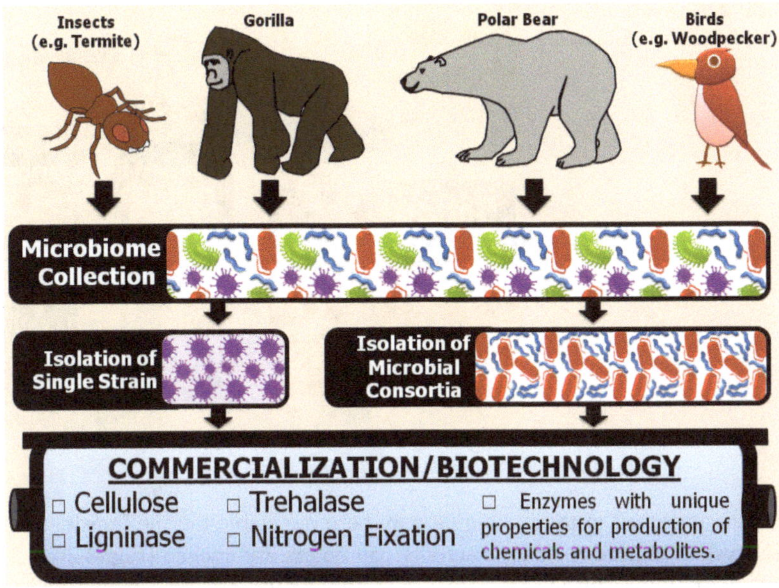

Fig. 4.23 Microbiome associated with other organisms can be mined for unique properties

organisms possess is indeed almost limitless. So next time you see yourself in the mirror or a person living in a remote part of the globe or you see a bird flying or a cow grazing, remember the invisible value buried within, they all beaming with miniature factories that are waiting to be unearthed. The invisible gold within us and other organisms is crying to ooze out (Fig. 4.23).

4.17 Tracking the Microbiome: A Security Windfall

Once the microbiome of each individual is mapped, it can serve as an important tool in the classification of people in different groups with a variety of other physiological and biomolecular descriptors. Each human can be tagged according to his or her microbial profile in manner akin to genetic or racial profiles. However, the exact identification will be relatively precise as one can be pinned down to one's unique microbial composition to the species level. Such identification pattern can also be useful in linking an individual to the company one is keeping and to the place one has been. The microbes from the surrounding say a farm will be part of the individual's microbiome if that person works in farm or has been there. An office worker in Canada will exposed to a different set of microbial landscape compared to someone working in an office in India. A person will be more readily pinned to a location or to an activity more readily than the techniques currently in place. The sampling can be as simple as collecting the air being exhaled or the skin microbes or microbial specimen from the clothing one is wearing. The core

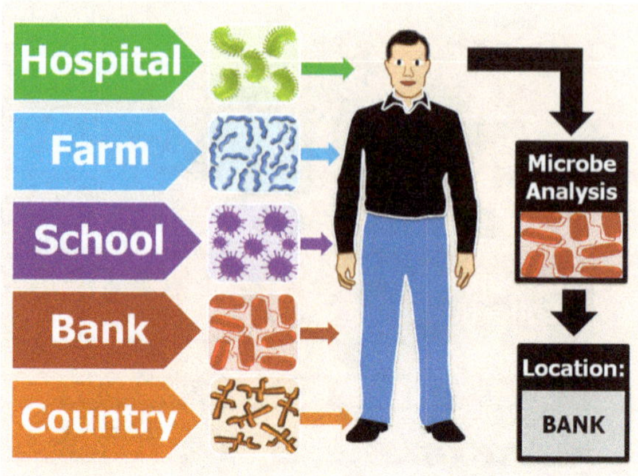

Fig. 4.24 A microbial registry for each location. An invisible identifier

microbiata attributed to the individual may be more stable but the transient microbial ecosystem indicative of their activity can be readily tracked (Fig. 4.24).

Our microbiome and the microbial communities in the environment and those of the people we are interacting with can intermingle more readily. The same cannot be said of the genetic profiles where the genetic material from the body components has to undergo very intimate contact in order to be exchanged. A cartographic depiction of the microbial fingerprint of places we frequent and of our homes will be make it easy to identify individual who have been to these locations. These microbial identifications can be catalogued, stored, retrieved and compared at will. Even the metabolites generated by microbial activity on the skin can be subject to microbiome policing. These metabolites will be indicative of the dermal microbes which in turn will be characteristic of the host. Intake of toxic material and drugs will reconfigure the microbial composition in the body. These in turn can be utilized to follow the biological changes one has been subjected. These biomarkers have not yet been fully exploited due to the dearth of information on the invisible organ. This is bound to change as the microbiome tracking will be an important diagnostic and law-enforcement tool. Homes can be uniquely marked with microbial components that will become part of anyone including an intruder who has been to the house. Microbial tracking will open up a completely new horizon for the security and surveillance industry.

4.18 Location and Activity Specific Microbes – An Intriguing Profiling Tool

Once the microbiome of all living organisms and the microbes associated with specific locations has been documented, the interactions amongst these invisible markers can be traced. These will permit the extraction of more pertinent

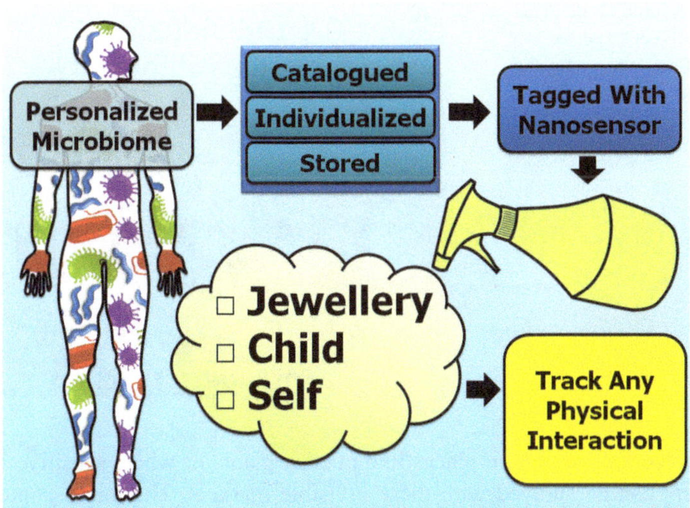

Fig. 4.25 Tagging valuables with individual-specific microbes

information about an individual and will help locate a person to a crime scene. Furthermore an accurate portrait of the presence of other microbe-harboring organisms or humans can be construed. This mode of detection will be akin to the identification of antibodies associated with people who have been exposed to viruses when travelling to different part of the globe. Remember the outbreak of the Ebola or Zika viruses where the link of people to such regions could only be assessed after the detection of the corresponding antibodies after certain incubation periods. In the case of the microbiome profiling, the presence of unfamiliar microbes can be culled from skin, mouth or exhaled air in a non-intrusive manner and analyzed. This microbe tracking technology can also be conceived for a variety of security-related applications. One's own unique microbe(s) can be sprayed on any valuables one possesses. If these microbe-marked belongings are stolen, they can easily be tracked to the individuals involved in the theft. On their arrest the presence of these uniquely tagged microbes will link the culprits to the stolen properties. Such evidence will be fool-proof. These nanotags can also be equipped with locating devices that can even reveal where the loot is safely nestled. This innovative technology based on an individual unique microbial ecosystem can help recover the stolen goods and can also facilitate the capture of the criminals. Furthermore, microbes can survive in an environment or on an object for a long time. Just image the economic benefit of this procedure that is almost infallible. Museums can apply this secure microbial brew on their priceless collections, banks can use them on their gold reserve and transfer of huge amounts of cash can occur safely or at least with the assurance that the microbially marked notes can be traced back readily. The stealth nature of this technology will make it also impossible to be neutralized and its specificity will hard to tamper with (Fig. 4.25).

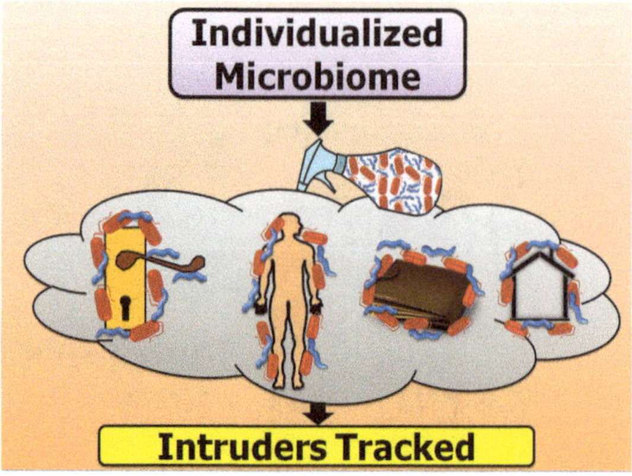

Fig. 4.26 Microbial idenfications (any object or person can be marked with a specific microbe that can be readily traced)

People working in secure places like nuclear plants or where sensitive information is kept can be sprayed with these invisible markers. The microscopic feature associated with this detection and tracking device will be so attractive that it will give other tracing technologies a real competition. Just imagine how safe one will feel wandering around wearing this invisible tracking device. One may even apply a touch of this microbial brew on one's kids with the comforting thought that if the children are involved in a foul play, they and the perpetrators can be traced. Women can apply this to their body and rest assured that if any malicious acts are inflicted on them, the culprits will be found. No DNA sample and evidence that can be destroyed easily are required. The offenders will be caught red-handed as these microbes will be unique to an individual. This new security landscape afforded by the microbiome-based technology will be a boon to the law-enforcement agencies and the justice systems around the globe. Even prisoners and violent offenders can be tracked with these invisible labels without their knowledge and all their goings and comings can be unceremoniously recorded, even to places where closed circuit camera may be impossible to be positioned. Indeed like any new technology, this device can fall in the wrong hands and its applications if not well-vetted may result in undesired outcomes (Fig. 4.26).

4.19 Conclusion

The presence of human microbiome and for that matter the microbial ecosystems of all organisms and their physiological contributing are fast becoming reality. This component of the human body is only now beginning to be unraveled. The invisible organ that spans all our external being, the digestive system, the lungs and the urogenital track is an integral part of the human identity. As any constituent of the body it performs a variety of tasks that cannot be executed without the input of these microbial communities that constitute the microbiome. Thus, it stands to reason that this long forgotten biological component needs to be actively studied and appreciated if we are to understand how the body works, falls sick and ages. The microbiome may be one of the last organ to be uncovered due to its invisibility and diverse location. The lack of tools has also contributed to this neglect. With ongoing progress in the NPS technology coupled with the bioinformatics based innovations, the obscurity that has marred this organ is fast dissipating. The ability to see the microbes in situ that is as they operate in communities is leading to a real appreciation of their role in the human body. Deciphering the dynamics of these microbial networks, the products they generate and the interactions they participate in, one can at last admire the amazing tasks they undertake on behalf of humans. The biological goods in form of nutrients and signaling molecules they synthesize in support of other organs of the body are critical for our survival. Most if not all visible organs have evolved to be dependent on this microbial assistance. How ineffective the gut will be without their invisible workers, otherwise it would be completely ineffective or entirely different anatomically. Understanding the minutiae of these nano-machines will help us optimize the operation of the body and will lead to technologies that will have far-reaching impact on the betterment of society. Remember, following the first visualization of microbes by Van Leuwenhoek, it took almost two centuries before a microbe could be cultured in the laboratory. Now one can scrutinize microbial communities without the necessity of culturing those that are unable to live independently.

If the unearthing of the invisible organ proceeds at a proper pace, in the near future the cure to a variety of maladies, and the creation of target-specific and side-effect averse medications will be within reach. More importantly health professionals will be able to restore physiological functions, modify behaviors and instill attributes that are at present time some people may discard as science fiction. The functional microbial communities will come not only from the unique diverse human population across the globe but will also be pooled from the vast reservoir of organisms roaming this planet, on land, in the ocean and also in the air. Some of select microbes that are not performing well will be renewed by the inoculation of targeted robust breed and by the intake of prebiotics or probiotics designed to reinvigorate the ailing communities. Microbial transplant is currently gaining some traction in hospitals and is perking the interest of the pharmaceutical industry. This will pick up momentum and microbiome-based therapy will be an important component of the health delivery system in not too distant a future. In the long term, it will be a go-to therapy for a number of diseases.

These novel technologies propelled by the discovery of the microbiome will inject major changes in the way health care is administered and in the regulations governing the delivery of wellness to people of advanced ages. This will trigger a dramatic transformation of health policies that will have significant impact on the health education and the pharmaceutical industry. Geriatric care will see a momentous shift as the microbial landscape in seniors experiences an earth-shattering modification in terms of diversity and abundance. The caring of the elderly will need to be re-invented with a microbiome-centric view as its driving force. Diseases provoked by dysbiosis will get a major therapeutic boost. The biotechnology industries will also be an important beneficiary of this newly acquired

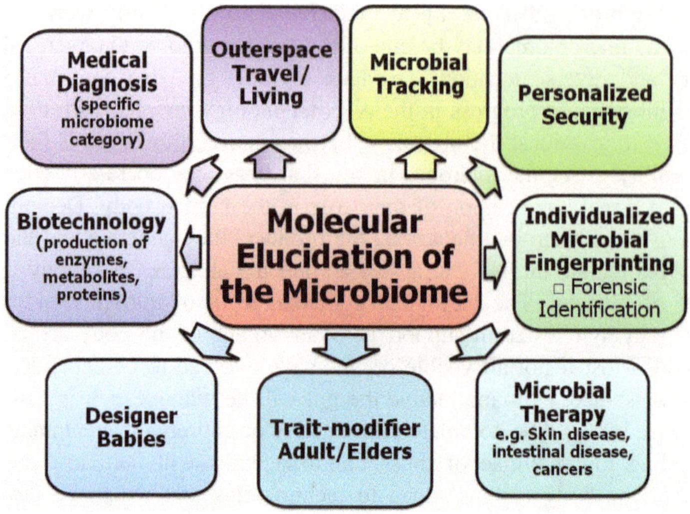

Fig. 4.27 A summary of the microbiome revolution and its associated technologies

microbial knowledge. Enlisting new microbes and microbial communities participating in the execution of a multitude of synthetic and physiological activities in their portfolio will open the door to unimaginable possibilities. Microbial pill or a prebiotic to enliven one's mood will be accessible. It will be almost like mood on the tap, all natural with virtually minimal side effect as there will be no chemical to metabolize or detoxify. Having pollution free fuels, precise drugs, physiology-modifying nutrients, a digestive system capable of degrading cellulose and miracle nano-devices to track valuables will be within the realm of possibility and reality. The future of microbiome has just begun. It will be terrific for human health and wellness. And beyond!!!!!!!!!!!!!! (Fig. 4.27).

Suggested Readings

Butel MJ (2014) Probiotics, gut microbiota and health. Med Mal Infect 44:1–8. https://doi.org/10.1016/j.medmal.2013.10.002

Cani PD, Van Hul M (2015) Novel opportunities for next-generation probiotics targeting metabolic syndrome. Curr Opin Biotechnol 32:21–27. https://doi.org/10.1016/j.copbio.2014.10.006

de Vos WM (2013) Fame and future of faecal transplantations–developing next-generation therapies with synthetic microbiomes. Microb Biotechnol 6:316–325. https://doi.org/10.1111/1751-7915.12047

Dietert RR, Dietert JM (2015) The microbiome and sustainable healthcare. Healthcare (Basel) 3:100–129. https://doi.org/10.3390/healthcare3010100

Gourbeyre P, Denery S, Bodinier M (2011) Probiotics, prebiotics, and synbiotics: impact on the gut immune system and allergic reactions. J Leukoc Biol 89:685–695. https://doi.org/10.1189/jlb.1109753

Jakobsdottir G, Nyman M, Fak F (2014) Designing future prebiotic fiber to target metabolic syndrome. Nutrition 30:497–502. https://doi.org/10.1016/j.nut.2013.08.013

Kali A (2015) Human microbiome engineering: the future and beyond. J Clin Diagn Res 9: DE01-04. https://doi.org/10.7860/JCDR/2015/14946.6570

Lloyd-Price J, Mahurkar A, Rahnavard G et al. (2017) Strains, functions and dynamics in the expanded Human Microbiome Project. Nature 550:61–66. https://doi.org/10.1038/nature23889

Luna RA, Foster JA (2015) Gut brain axis: diet microbiota interactions and implications for modulation of anxiety and depression. Curr Opin Biotechnol 32:35–41. https://doi.org/10.1016/j.copbio.2014.10.007

Marsh PD, Head DA, Devine DA (2014) Prospects of oral disease control in the future - an opinion. J Oral Microbiol 6:26176. https://doi.org/10.3402/jom.v6.26176

Miloh T (2015) Probiotics in pediatric liver disease. J Clin Gastroenterol 49(Suppl 1):S33–S36. https://doi.org/10.1097/MCG.0000000000000365.

Nunes-Alves C (2016) Microbiome: drugs for your bugs. Nat Rev Microbiol 14:62. https://doi.org/10.1038/nrmicro.2015.23

Serban DE (2014) Gastrointestinal cancers: influence of gut microbiota, probiotics and prebiotics. Cancer Lett 345:258–270. https://doi.org/10.1016/j.canlet.2013.08.013

Tomar SK, Anand S, Sharma P, Sangwan V, Mandal S (2015) Role of probiotics, prebiotics, synbiotics and postbiotics in inhibition of pathogens. in *The Battle Against Microbial Pathogens: Basic Science, Technological Advances and Educational Programs. Formatex.* Badajoz, Spain

Umu OC, Oostindjer M, Pope PB et al. (2013) Potential applications of gut microbiota to control human physiology. Antonie Van Leeuwenhoek 104:609–618. https://doi.org/10.1007/s10482-013-0008-0

Viaud S, Daillere R, Boneca IG et al. (2014) Harnessing the intestinal microbiome for optimal therapeutic immunomodulation. Cancer Res 74:4217–4221. https://doi.org/10.1158/0008-5472.CAN-14-0987

Glossary

Adrishta A word meaning invisible force in Sanskrit Language.

Acne A common skin disease when pores are clogged.

Actinobacteria A phylum of Gram-positive bacteria.

ATP Adenosine triphosphate, the energy currency in living organisms.

Amylase An enzyme that converts starch and glycogen into glucose.

Alginate A polysaccharide produced by some bacteria prominent during lung infections.

Akkermensia muciniphilia A bacterium that resides in human intestine and may be involved in obesity and other diseases.

Androstenol A male pheromone that can cause attraction to women.

Androstenone A pheromone found in urine and axillary sweat in humans.

Animalcule A microscopic animal.

Anthrax An infection caused by the bacterium *Bacillus anthracis*

Anti-proteases Molecules designed to inhibit the activity of proteases.

Apigenin A natural plant product with numerous health benefits.

Apoptosis A process of programmed cell death.

Archaea Microbes resembling bacteria with unique chemical features.

Aspergillus oryzae A fungus found in fermented foods rich in enzymes like amylase and lipases.

Atopobium spp. Bacteria associated with vaginosis and other diseases.

Atopic dermatitis A type of eczema when the skin becomes red and itchy.

Azoreductase An enzyme that can activate and de-activate drugs.

Bacterium Single-cell organism lacking organelles.

Bacteriodes Anaerobic bacteria also found in humans.

Bacteriodes fragilis A bacterium associated with good health in Malawian children.

BC Before Christ.

Bifidobacterium breve A bacterium with wide ranging health benefits.

Bifidobacterium fragilis A bacterium in the gut that helps fortify the immune system.

V.D. Appanna, *Human Microbes - The Power Within*,
https://doi.org/10.1007/978-981-10-7684-8

Bioinformatics A subject that utilizes computational tools to analyze biological data.

Bile acid An acid produced by the liver and is involved in the digestion of fatty acids.

Bisphenol A A chemical utilized in the synthesis of plastic that interferes with the endocrine system.

2,3 Biphosphoglycerate A metabolite involved in the release of oxygen from haemoglobin.

Biofilms Exopolysaccharides produced by microbes for a variety of functions.

Body Mass Index BMI The measurement of body fat based on height and weight.

Butanoic acid A short-chain fatty acid with an unpleasant smell.

Cadaverine A polyamine responsible for malodour.

Cadmium A metal pollutant that leads to dysbiosis

Carbohydrates Biomolecules essential for energy production.

Carboxymethyl cellulose (CMC) A food additive derived from cellulose.

Carnitine A naturally occurring amino acid involved in the metabolism of fats.

Catecholamine Catechol containing molecules that are hormones or neurotransmitters.

Cefoxilin An antibiotic that blocks the synthesis of bacterial cell wall.

Cellulose A polymer of β-glucose found mostly in plant products.

Chitin A biomolecule containing N-acetylglucosamine.

Chitinase An enzyme that breaks chitin into N-acetylglucosamine.

Chromosome Nuclear material where genes are stored.

Clostridium difficile A bacterium involved in a variety of diseases that may become resistant to antibiotics.

Clostridium symbiosis A bacterium that helps in cellulose metabolism and can reverse growth deficit.

Chlorhexidine A disinfectant found in mouthwash.

COPD A chronic obstructive pulmonary disease characterized by breathlessness.

Creatine A nitrogen containing compound involved in ATP (energy) homeostasis.

Crohn's Disease A disease of the bowel characterized by abdominal pain.

Curcumin A plant product with anti-oxidant and anti-inflammatory properties.

Cystic fibrosis A genetic disease that primarily affects the lungs.

Dahi A fermented milk product originating from India.

Deoxyribonucleic acid DNA The chemical carrying genetic information.

Desulfovibrio spp. A bacterium associated with increased consumption of alcohols and preserved foods.

Dermicidin An anti-microbial peptide secreted by human sweat glands.

Dopamine A chemical secreted by the brain cells.

Dosa A pancake made from fermented batter of lentils.

Dysbiosis An imbalance of microbial system in the human body.

Ebola virus A virus causing severe hemorrhagic fever.

Escherichia coli Nissle A bacterium used as a probiotic for a variety of gastrointestinal ailments.

Enzymes Biomolecules that catalyze biochemical reactions and are mostly proteins.

Enterobacteria spp. A bacterium that is abundant during irritable bowel syndrome.

Enterotype A classification system based on the bacteria residing in the human gut.

Erythropoietin EPO A hormone produced in the kidney that promotes the formation of red blood cells.

Ethyl phenyl sulphate A chemical of microbial origin that may be associated with autism.

Eubacterium ramulus A gut microbe that can metabolize flavonoids and may lead to weight gain.

Eukaryote An organism with a nucleus.

Fatty liver disease A disease of the liver resulting in the accumulation of triglycerides.

Firmicutes A phylum of usually rod-shaped bacteria.

Filaggrin A protein involved in skin maturation.

Flavanoids A group of plant-derived polyphenol containing compounds with numerous health benefits.

FOS Fucosylated oligosaccharides (prebiotics) usually found in milk that favour the growth of beneficial bacteria.

Fucosylated The process of adding fucose, a hexose deoxy monosaccharide.

Fungus A single-cell or multinucleate organism.

Fusobacteria A family of slender rod-shaped bacteria.

Galactosidase An enzyme that breaks chemical bonds made up of galactose.

GABA γ Amino butyric acid is a neurotransmitter that reduces neuronal activity.

Gastric juice A colourless acidic liquid secreted by the stomach.

Glucagon like peptide (GLP) A biomolecule that works like the hypoglycemic hormone glucagon.

Glutathione transferase An enzyme that adds glutathione to other molecules.

Gene Contains DNA and carries instruction to make proteins.

Genome All the genetic material in an organism.

Gnotobiotic An environment without microbes.

GOS Galactose-containing oligosaccharides that are prebiotics.

Granulicella A genus of bacteria abundant in individuals with dental cavities.

Halitosis A disease characterized by bad breath.

HDAC Histone deacetlylases are a class of enzymes that regulate nuclear metabolism.

HDL High density lipoprotein, a marker of healthy cholesterol.

Helicobacter pylori A bacterium that can infect the stomach.

Hydrogen sulphide A gas that helps in blood flow in the arteries.

HMP Human Microbiome Project is an initiative to identify all the microbes found in humans.

Hydrothermal Vent A fissure with geo-thermally heated water.

Hyperoxaluria A disease characterized by high oxalate levels in the urine.

Hypothiocyanite A natural anti-microbial.

Homocysteine An amino acid found in high amounts during some diseases.

Interleukin-12 A natural protein involved in immune response.
Inulin A dietary fibre containing fructans.
Irritable bowel syndrome A common disease affecting the large intestine.
Kefir A fermented milk beverage.
Kimchi A fermented Korean staple made of cabbage and other vegetables.
Lactobacillus acidophilus A bacterium that bestows numerous health benefits.
Lactobacillus casei A bacterium known to decrease symptoms associated with rheumatoid arthritis.
Lactobacillus rhamnonus A bacterium known to improve defecation frequency.
Lactoferrin A protein found in milk that has antimicrobial property.
Lactoperoxidase An enzyme that has anti-microbial property.
Lactase An enzyme involved in the digestion of lactose, the sugar in milk.
LDL Low density lipoprotein, a marker for unhealthy cholesterol.
Leptin A hormone synthesized by adipose tissues that is involved in energy balance.
Leucine An essential amino acid for humans.
Lignin A phenolic polymer found in plants.
Lima Syndrome A feeling of affection of kidnappers for their victims.
Linoleic acid A polyunsaturated omega fatty acid with numerous health benefits.
Lipase An enzyme that breaks bonds involving fatty acids.
Lipids Biomolecules containing or derived from fatty acids.
Lipoic acid An organosulfur containing chemical with numerous biological properties.
Lipopolysaccharide A chemical containing both lipid and polysaccharide.
Lucerin A light-emitting compound found in living organisms.
Lymphoma Cancer of the lymph nodes.
Mawe A fermented maize dough.
Meconium The first stool of human infant.
Melanin Natural pigments found in the skin.
Mercury A metal pollutant that promotes microbial imbalance in the body.
Meta organisms The disparate biological entities residing within a living system.
Metabolites Products formed and utilized during metabolism.
Metabolome All the metabolites in a living organism.
Metagenome All genetic material from a natural sample in the environment.
Metformin A medication for type II diabetes that may impede the mobility of the gut.
Micro-array A laboratory technique to detect DNA and RNA.
Microbe Single-cell organism invisible to the naked eyes.
Microbiome All microbes within and on an organism.
Microbiota Microbes inhabiting an ecosystem.
Miso A Japanese seasoning obtained by fermenting soya beans.
***Moraxella* spp.** A genus of bacteria found in nasal cavities and lungs.
Mucin Glycoproteins found in mucus.
NGS Next generation sequencing involves the modern techniques to sequence DNA and identify organisms in their natural environments.

Nanotag A molecular identifying label.
Necrotizing enterocolitis A disease when damaged intestinal tissue begins to die.
Nitric oxide A biological gas involved in numerous bodily functions.
Nitrogen Fixers Organisms that can convert nitrogen in the air into ammonia.
Nitroreductases Enzymes involved in the reduction of nitrogen containing compounds.
Nori Japanese name for edible seaweeds.
Nucleus Where all genetic material in a eukaryotic organism is stored.
Oligosaccharides Molecules containing a few monosaccharides.
Organogenesis A process during which organs develop.
Oxytocin A biomolecule linked to humour and laughter.
PCR Polymerase Chain Reaction is a technique used to make copies of genes.
Paleo-diet A diet consisting of foods like fruits, nuts, roots, vegetables that was available during paleolithic era.
Parkinson's disease A degenerative disease that affects the motor system.
Pasteurization A process of killing microbes in foods discovered by the French scientist Louis Pasteur.
Psychobiotics Probiotics or prebiotics that can help improve mental health by modifying the microbiome in the body.
Peptidoglycan A biomolecule containing amino acids and monosaccaharides found in microbial cell walls.
Peptoniphilus A genus of Gram-positive bacteria that increases during menopause.
Prevotella A genus of Gram-negative bacteria associated with the oral and vaginal microbial flora.
Propionobacterium acnes A rod-shaped Gram positive bacterium linked to acne.
PUFA Polyunsaturated fatty acids contain more than one double bonds and are essential for good health.
Pilosebaceous A microscopic exocrine gland in the skin that secretes sebum.
Phylum A primary category in the classification of living organisms.
Polyphenol A component of numerous foods that has anti-oxidant properties.
Polysaccharides Molecules containing numerous simple monosaccharides like glucose.
Porphyran A sulphate containing carbohydrate found in seaweeds.
Porphyromonas gingivalis A bacterium responsible for the destruction of gums.
Porphyrin A molecule containing pyrrole found in haemoglobin.
Prebiotics Nutrients that promote the growth of healthy bacteria in the body.
Predinosone A synthetic steroid utilized to treat numerous diseases.
Probiotics Live microbes with health benefits.
Prokaryote Single-celled organism without a nucleus.
Prevotella copri A bacterium that is abundant during rheumatoid arthritis.
Proteases Enzymes that degrade proteins.
Proteins Biomolecules containing nitrogen that are essential nutrients for humans.
Proteobacteria A phylum of bacteria found in humans.
Proteome All the proteins expressed in a living organism.
Psoriasis An autoimmune disease characterized by patches of abnormal skin.

Rebiosis A procedure to re-establish microbial balance in the body.

Rhesus factor An antigen found in the red blood cells.

Rheumatoid arthritis An autoimmune disease resulting in chronic inflammation of the joints.

Ribonucleic acid RNA The chemical that decodes genetic information.

Resvesatrol A phenolic compound found in numerous plants including grapes and berries linked to healthy outcomes.

Rosburia spp. A genus of anaerobic bacteria in the gut that may be associated with healthy outcomes.

Rumen The first chamber of the alimentary canal of ruminants that serves as the primary site of microbial fermentation.

Ruminococcus gnavus An intestinal bacterium that can metabolize mucin.

16S rRNA 16 S ribosomal ribonucleic acid gene is utilized to identify and classify bacteria.

Sake A Japanese rice wine.

Sialic acid A chemical found in human milk associated with good health in infants.

Sebum The oily secretion of the skin.

Serotonin A neurotransmitter associated with mood and happiness.

Shoyu A dark brown liquid obtained from fermented soya beans.

Skin cartography Metabolites and bacteria on the skin indicative of a specific individual.

Sorivudine A thymine analogue serving as anti-viral drugs.

Species A group of organisms with similar genetic information that they can exchange.

Starch A polymer of α glucose and an important source of energy in humans.

Stockholm Syndrome A feeling of affection of the victims towards their kidnappers.

Strain A genetic variant of a microorganism.

Staphylococcus aureus A bacterium that increases during atopic epidermis.

Staphylococcus epidermis A bacterium being introduced in cream to fight dermal dysbiosis.

Staphylococcus horminis A bacterium associated with the worst odour in humans.

Streptococcus A genus of bacteria associated with a variety of infections.

Supraorganism Collection of different organisms that behave like a single entity.

Synbiotics A mixture of prebiotics and probiotics given to restore healthy microbes in the body.

α Synuclein A neuronal protein involved in Parkinson's disease.

Thioalcohol An alcohol containing sulphur associated with strong odours.

Triclosan A chemical added to a variety of consumer products in order to reduce bacterial contamination.

Trimethylamine oxide (TMAO) A metabolite produced by gut microbes and linked to various diseases.

Treponema A genus of spiral-shaped bacteria found in rural communities in Africa and is linked to good health.

Tryptophan An essential amino acid.

Ulcerative colitis A disease that causes inflammation in the large intestine.

Uncoupling protein (UCP) A protein that dissipates the proton gradient and is responsible to make energy in the mitochondria.

Ugi A porridge made of corn and millet.

Urate A chemical derived from nucleic acid metabolism that can result in gout.

Trait Characteristic belonging typically to a specific organism or individual.

Vaginosis A disease of the vagina caused by microbes.

Verrucomicrobacteria A phylum of wart-shaped bacteria found in human.

Virus Single-cell organism that needs other organisms to live.

Vitamin B An essential nutrient involved in metabolism.

Vitamin K An essential nutrient involved in blood coagulation.

XOS Xylose containing oligosaccharides that are probiotics.

Yakult A Japanese milk drink rich in probiotics.

Zika virus A virus that causes the Zika fever, a disease associated with red eyes and joint pain.

Index

V.D. Appanna, *Human Microbes - The Power Within*,
https://doi.org/10.1007/978-981-10-7684-8

FSC
www.fsc.org
MIX
Papier | Fördert
gute Waldnutzung
FSC® C083411

Kolibri 360 GmbH
Ferdinand-Jühlke-Straße 7
99095 Erfurt, Deutschland
produktsicherheit@kolibri360.de